Mastering Swift 6

Seventh Edition

Modern programming techniques for high-performance apps in Swift 6.2

Jon Hoffman

‹packt›

Mastering Swift 6

Seventh Edition

Copyright © 2025 Packt Publishing

Portfolio Director: Ashwin Nair
Relationship Lead: Suman Sen
Project Manager: Ruvika Rao
Content Engineer: Nithya Sadanandan
Technical Editor: Sweety Pagaria
Copy Editor: Safis Editing
Indexer: Rekha Nair
Proofreader: Nithya Sadanandan
Production Designer: Ajay Patule
Growth Lead: Sohini Ghosh

First published: June 2015
Second edition: November 2015
Third edition: October 2016
Fourth edition: September 2017
Fifth edition: April 2019
Sixth edition: November 2020
Seventh edition: September 2025

Production reference: 1080825

Published by Packt Publishing Ltd.
Grosvenor House
11 St Paul's Square
Birmingham
B3 1RB, UK.
ISBN 978-1-83620-369-8

www.packtpub.com

With this being the seventh book in the Mastering Swift series, I would like to thank everyone who has offered encouragement, positive feedback, and constructive criticism over the years. This includes my family, friends, and everyone at Packt, especially three very special people who inspire me and make life so wonderful: my two amazing children, Kailey and Kai, who make every day interesting and always make me proud; and my fiancée, Heidi, whose love and support mean the world to me and with whom I finally see a future filled with love and shared adventures.

- Jon Hoffman

Contributors

About the author

Jon Hoffman has over 30 years of experience in the information technology field. Over the years, he has worked in system administration, network administration and security, application development, and architecture. He currently serves as an Enterprise Software Manager for Syntech Systems.

Outside of his professional life, Jon has a wide range of personal interests that keep him both physically and mentally engaged. He enjoys spending quality time with his two children and his fiancée. He also stays active through running, hiking, paddleboarding, yoga, and working out. In addition, Jon has a deep passion for reading and continues to nurture his love for coding.

About the reviewer

Juan C. Catalan is a software engineer with more than twenty years of professional experience. He began mobile development in the early days of iOS 3 and has since worked as a professional iOS developer across a variety of industries, including medical devices, financial services, real estate, document management, fleet tracking, and industrial automation. Juan has contributed to over thirty apps published on the App Store, several of which serve millions of users.

He gives back to the iOS community through technical talks, developer mentoring, book reviews, and—since 2023—as the author of *SwiftUI Cookbook*, *Third Edition*, published by Packt. Juan lives in Austin, Texas, with his wife, Donna, and their children.

Emil Atanasov is an IT consultant with extensive experience in mobile technologies. He began working in mobile development in 2006 and now runs his own contracting and consulting company, Appose Studio Inc., serving clients around the world.

He holds an MSc from RWTH Aachen University in Germany and Sofia University "St. Kliment Ohridski" in Bulgaria. Emil has contracted for several large companies in the U.S. and U.K., serving as a software architect, project manager, iOS developer, and Android developer. He also teaches courses at Sofia University in Swift and iOS Development with SwiftUI and is the author of *Learn Swift by Building Applications*.

Mark Kuharich has over twenty years of experience in the engineering and IT fields. He earned a bachelor's degree in computer science from the United States Military Academy at West Point, New York. Mark has worked on the Starbucks mobile app, served as an iOS developer for the Disney Parks and Resorts app, and was part of the Siri team at Apple.

In addition to his professional achievements, Mark is passionate about playing golf.

I would like to thank the author, Jon Hoffman; the Packt team; and my wife—without whom none of this would be possible.

Stewart Lynch is a retired educator and IT professional who began programming in 1969 on an IBM 360 computer in FORTRAN while earning a degree in mathematics.

Stewart spent over thirty years in the education sector as a teacher, administrator, and IT director in two large school districts in British Columbia, Canada, before taking early retirement to work for Canada's largest software company.

Throughout his career in education, Stewart pursued his passion for coding, developing software solutions for both staff and students. This passion carried into his work in the public sector, where he taught coding to customers and created value-added software solutions for clients worldwide.

Stewart is almost entirely a self-taught software developer, committed to staying current with the latest developments in iOS and macOS with Swift and SwiftUI. He has multiple apps on the App Store but is now focused on helping others learn and improve their coding skills through his YouTube channel (youtube.com/@StewartLynch).

I would like to thank the author of this book, Jon Hoffman, for being so responsive and for quickly answering my questions throughout this process.

Zhanserik Suyunbayev is an experienced iOS software engineer with over 8 years of expertise in native iOS development. He holds a master's degree in Applied Computer Science from Saint Petersburg State University of Engineering and Economics. Most recently, he worked on a project for Bank of America, where he was responsible for developing and maintaining iOS applications using technologies such as Swift, SwiftUI, Objective-C, Xcode, Auto Layout, REST APIs, and Git. Zhanserik has also contributed to high-quality, reliable iOS solutions for other prestigious clients, including Amazon.com Inc., Citibank, and MoneyGram Inc.

Table of Contents

Chapter 10: Custom Subscripting   147

Chapter 11: Property Observers and Wrappers   161

Chapter 14: Structured Concurrency 203

Chapter 15: Memory Management 229

Chapter 16: Advanced and Custom Operators 243

Chapter 17: Access Controls 263

Preface

Since its surprise debut at WWDC in June 2014, Swift has rapidly redefined how we build software, combining safety, speed, and a clear, concise syntax to power everything from iOS, macOS, watchOS, and tvOS apps to server-side services on Linux and, since September 2020, native Windows development. Born from Chris Lattner's vision of a modern successor to Objective-C, Swift's modern features, such as optionals, generics, and closures, combined with its powerful type inference system, have made code more expressive, reliable, and enjoyable to write. While Apple's decision to open-source the language in 2015 and stabilize its ABI in Swift 5 has created a vibrant community around swift.org. In this book, we'll explore some of Swift's most advanced features to help you take your development to the next level, writing cleaner, more powerful, and highly optimized applications.

Who this book is for

This book is written for developers with an understanding of the Swift programming language who are looking to elevate their skills by exploring advanced topics and techniques. The examples provided are compatible with application development across all Apple platforms, including macOS, iOS, iPadOS, visionOS, and watchOS, as well as Linux and Windows development, unless otherwise specified.

What this book covers

Chapter 1, The Evolution of Swift, highlights the evolution of the Swift language, showing how its history shaped the features and design choices that developers use today. It also introduces swift.org, a key resource providing documentation, collaboration opportunities, and support for the Swift community.

Chapter 2, Closures and Result Builders, introduces closures, which are self-contained blocks of code that can capture and retain references to variables from their surrounding context. Additionally, we will look at result builders, a feature that allows developers to create custom **Domain-Specific Languages (DSLs)** used for defining complex data structures.

Chapter 3, Protocols and Protocol Extensions, explores protocols, highlighting how they can be used as full-fledged types that enable polymorphism through the unified interface they provide. It also introduces protocol extensions, which help reduce code duplication by offering default implementations for methods and properties. Additionally, the chapter covers the use of Any to represent an instance of any type, and *existential any*, which allows storing any value that conforms to a specific protocol.

Chapter 4, Generics, explores generic types, which serve as the foundation for many types in the Swift standard library. Real-world examples demonstrate how to create and use generics effectively. Additionally, the chapter covers topics such as generic subscripts and associated types, illustrating how generics can enhance the flexibility of our code.

Chapter 5, Value and Reference Types, explains the differences between value types and reference types in Swift. This chapter includes practical examples of structures and classes, covering features such as copy-on-write for optimizing performance with large value types and recursive data types for reference types. Additionally, we will look at dynamic dispatch, which, while introducing some performance overhead, enables the flexibility needed for class hierarchies.

Chapter 6, Enumerations, looks at Swift enumerations and what makes them more powerful than enumerations in other languages. We will also look at how to use raw values and how pattern matching can simplify our code. Additionally, we will see how associated values allow Swift enumerations to represent complex data structures.

Chapter 7, Reflection, looks at what reflection is and how the Mirror API in Swift enables it. We will demonstrate how to use the Mirror API to inspect structures and classes, revealing details such as property names, values, and type information.

Chapter 8, Error Handling and Availability, looks at Swift's error handling mechanisms and how errors are defined using types conforming to the Error protocol. We will demonstrate how to throw errors in functions using the `throws` keyword and handle them with `do-catch` blocks. Additionally, we will discuss Swift's `availability` and `unavailability` attributes for conditional code execution based on platform versions.

Chapter 9, Regular Expressions, explores the power of regular expressions for pattern matching and text manipulation, from simple searches to complex data extraction. The chapter shows how Swift supports regular expressions with regular expression literals and the `Regex` type. Additionally, `Regex Builder` is introduced, which offers a clearer, more intuitive way to define patterns, and we will look into advanced features such as transforming and capturing matches using references.

Chapter 10, Custom Subscripting, explores how incorporating subscripts into custom types can improve readability and usability. We will examine how to create multi-parameter subscripts and discuss how to use subscripts appropriately, consistent with the Swift language itself.

Chapter 11, Property Observers and Wrappers, explores how property observers and wrappers automatically respond to changes in property values, which improves the responsiveness of our applications. We will examine how property observers trigger actions while property wrappers abstract property management into reuseable types.

Chapter 12, Dynamic Member Lookup and Key Paths, explores how dynamic member lookups allow properties to be resolved at runtime. Additionally, we will examine how key paths offer a type-safe way to access and manipulate properties using the \. syntax.

Chapter 13, Grand Central Dispatch, explores the difference between concurrency and parallelism. We will see how Grand Central Dispatch uses both serial and concurrent queues to handle task execution. Additionally, we will highlight how it has tools for managing UI updates and scheduling tasks effectively.

Chapter 14, Structured Concurrency, explores how async and await offer intuitive ways to handle asynchronous tasks, allowing functions to pause and resume without blocking the main thread. We will examine how tasks and task groups further control and coordinate asynchronous operations. Additionally, we will look at actors and the role they play in managing state safety.

Chapter 15, Memory Management, examines how Swift uses **Automatic Reference Counting (ARC)** to manage memory for reference types such as classes. We will discuss how to use weak and unowned references to avoid strong reference cycles, which can hinder ARC from deallocating reference types, causing memory leaks.

Chapter 16, Advanced and Custom Operators, explores advanced bitwise operators, such as the AND, OR, XOR, and NOT operators, for manipulating variable bits and the right and left shift operators for shifting bits. We will also look at how to add operator methods to our custom types.

Chapter 17, Access Controls, looks at how access controls are used to ensure code security by restricting access to specific parts of the code base. We will examine the five access levels that Swift provides and how to apply them effectively. Additionally, we will see how applying the principle of least privilege will ensure that only the necessary parts of our code are accessible.

Chapter 18, Swift Testing, looks at Swift Testing, which is a powerful new testing framework introduced in Swift 6. We will explore the key building blocks, such as the @Test attribute and the #expect and #require macros, as well as the use of traits for adding metadata.

Chapter 19, Object-Oriented Programming with Swift, looks at object-oriented programming and how its design principles can be effectively applied with Swift to create structured and reusable code.

Chapter 20, Protocol-Oriented Programming with Swift, explores the core principles of protocol-oriented programming with Swift and how it differs from traditional object-oriented programming. Additionally, we will examine how Swift's standard library is built with a protocol-oriented approach.

Chapter 21, Functional Programming with Swift, looks at Swift's support for functional programming concepts such as immutability, pure functions, and first-class functions. We will also explore advanced techniques such as function composition, currying, and recursion to manage complex operations effectively.

To get the most out of this book

To get the most out of this book, you will need a basic understanding of the Swift language and a basic understanding of modern development techniques. All code examples have been tested using Xcode 26 on a Mac; however, all examples should also work using Swift on Linux or Windows.

Download the example code files

The code bundle for the book is hosted on GitHub at https://github.com/PacktPublishing/Mastering-Swift-6-Seventh-Edition. We also have other code bundles from our rich catalog of books and videos available at https://github.com/PacktPublishing. Check them out!

Conventions used

There are a number of text conventions used throughout this book.

CodeInText: Indicates code words in text, database table names, folder names, filenames, file extensions, pathnames, dummy URLs, user input, and Twitter handles. For example: "Notice that we added the any keyword in the parameter definition."

A block of code is set as follows:

```
func drawAll(_ items: [any Drawable]) {
    for item in items {
        item.draw()
    }
}
```

When we wish to draw your attention to a particular part of a code block, the relevant lines or items are set in bold:

```
enum Weather {
    case sunny
    case cloudy
    case rainy(Int)
    case snowy(amount: Int)
}
```

Bold: Indicates a new term, an important word, or words that you see on the screen. For instance, words in menus or dialog boxes appear in the text like this. For example: "To use Xcode's built-in conversion tool, simply right-click on the regular expression you would like to use, select **Refactor**, and then select **Convert to Regex Builder**."

Warnings or important notes appear like this.

Tips and tricks appear like this.

Get in touch

Feedback from our readers is always welcome.

General feedback: If you have questions about any aspect of this book or have any general feedback, please email us at customercare@packt.com and mention the book's title in the subject of your message.

Errata: Although we have taken every care to ensure the accuracy of our content, mistakes do happen. If you have found a mistake in this book, we would be grateful if you reported this to us. Please visit `http://www.packt.com/submit-errata`, click **Submit Errata**, and fill in the form.

Piracy: If you come across any illegal copies of our works in any form on the internet, we would be grateful if you would provide us with the location address or website name. Please contact us at `copyright@packt.com` with a link to the material.

If you are interested in becoming an author: If there is a topic that you have expertise in and you are interested in either writing or contributing to a book, please visit `http://authors.packt.com/`.

Share your thoughts

Once you've read *Mastering Swift 6, Seventh Edition*, we'd love to hear your thoughts! Scan the QR code below to go straight to the Amazon review page for this book and share your feedback.

`https://packt.link/r/1836203691`

Your review is important to us and the tech community and will help us make sure we're delivering excellent quality content.

1

The Evolution of Swift

Swift, which was introduced by Apple in 2014, is a versatile and modern programming language designed for building applications across Apple platforms such as iOS, macOS, watchOS, visionOS, and tvOS. However, it can also be used for server-side Linux development, Windows development, and even embedded systems. In September 2020, Apple officially released a version of Swift that can be used for Windows development.

Swift's seamless blend of safety, speed, and ease of use has made it an attractive choice for developers of all levels, offering a unified language experience across the various platforms. Its concise syntax helps make code clear and easy to understand, while its type inference system helps catch mistakes early on, making code more reliable. Additionally, Swift introduces modern programming concepts such as optionals, generics, and closures, enabling developers to write cleaner and more flexible code. With support from a vibrant community and regular updates from Apple, Swift continues to evolve, providing a user-friendly platform for creating applications across different devices and operating systems.

In this chapter, you will learn about the following topics:

- How Swift has evolved
- Migrating existing projects to Swift 6
- How swift.org can help you

Before we get started with Swift, let's look at how you can get the most of this book.

Getting the most out of this book — get to know your free benefits

Unlock exclusive **free** benefits that come with your purchase, thoughtfully crafted to supercharge your learning journey and help you learn without limits.

Here's a quick overview of what you get with this book:

Next-gen reader

Our web-based reader, designed to help you learn effectively, comes with the following features:

⟲ Multi-device progress sync: Learn from any device with seamless progress sync.

🖹 Highlighting and notetaking: Turn your reading into lasting knowledge.

🔖 Bookmarking: Revisit your most important learnings anytime.

🔅 Dark mode: Focus with minimal eye strain by switching to dark or sepia mode.

Figure 1.1: Illustration of the next-gen Packt Reader's features

Interactive AI assistant (beta)

Our interactive AI assistant has been trained on the content of this book, to maximize your learning experience. It comes with the following features:

✦ Summarize it: Summarize key sections or an entire chapter.

✦ AI code explainers: In the next-gen Packt Reader, click the Explain button above each code block for AI-powered code explanations.

Note: The AI assistant is part of next-gen Packt Reader and is still in beta.

Figure 1.2: Illustration of Packt's AI assistant

DRM-free PDF or ePub version

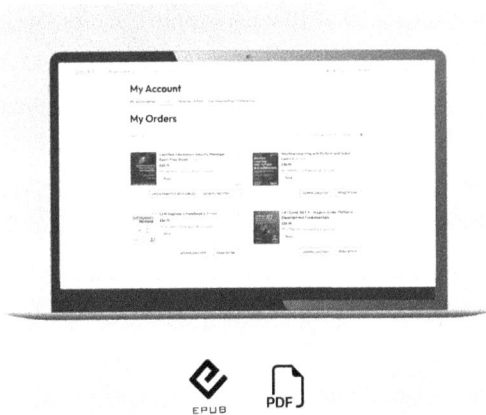

Learn without limits with the following perks included with your purchase:

📄 Learn from anywhere with a DRM-free PDF copy of this book.

📄 Use your favorite e-reader to learn using a DRM-free ePub version of this book.

Figure 1.3: Free PDF and ePub

Now let's begin our journey by looking at the evolution of the Swift language.

How Swift has evolved

Swift made its debut at the **Worldwide Developers Conference** (WWDC) in June 2014, surprising many in the tech industry. Swift was arguably the most significant announcement at WWDC 2014 and very few people, including Apple insiders, were aware of the project prior to it being announced.

Before we start to look at the language itself, let's see how it has evolved.

Swift was born

Development of the Swift language itself can be traced back to 2010 when Apple's Senior Director of Developer Tools, Chris Lattner, started designing the basic concepts for a new language. At the time, Apple was heavily reliant on Objective-C, which had been their primary language since the 1980s. While Objective-C had served Apple well, the company recognized the need for a more modern and safer language, but also the need for a language that could seamlessly interoperate with existing Objective-C code, allowing developers to adopt Swift incrementally without rewriting entire codebases.

The introduction of Swift wasn't just another product launch; it marked a significant shift in Apple's programming landscape. The announcement demonstrated Apple's commitment to pushing boundaries and redefining how software for Apple's devices was written. In addition, it demonstrated Apple's commitment to the developer community, which has only grown stronger as Swift has evolved.

Swift evolved

At WWDC 2015, Apple continued the evolution of Swift by announcing Swift 2, which brought substantial improvements to the language. What was particularly noteworthy was that these enhancements were based on feedback directly from the developer community. This highlighted Apple's willingness to listen and adapt, fostering a more collaborative environment with the developer community.

However, the most significant announcement of WWDC 2015 was arguably the decision to make Swift an open-source project. This move not only increased access to the language but also encouraged collaboration and innovation on a broader scale. It also represented a departure from Apple's traditional closed-door approach and a move to a more inclusive future of software development.

Four years later, in 2019, Swift 5.0 was officially released, introducing a stable version of the **application binary interface (ABI)** across Apple's platforms. This was a major step forward because it addressed a longstanding challenge of binary compatibility, enabling Swift code compiled with one version of the compiler to interact with code compiled with future versions. This ensured compatibility and stability as Swift evolved. Developers could now confidently distribute precompiled libraries, knowing they would remain compatible with future versions of Swift. This binary compatibility is crucial for the long-term sustainability of the language.

The stable ABI reinforced Swift's position as a versatile language for building cross-platform applications across Apple's devices. With binary compatibility now guaranteed across different platforms, developers could leverage Swift to target a broader range of devices and operating systems, including iOS, macOS, watchOS, visionOS, and tvOS. This cross-platform capability enabled developers to write code once and deploy it across the various platforms, streamlining the development process.

The Swift 5 release, coupled with the stable ABI, was a crucial moment in Swift's evolution, providing developers with a stable foundation for building high-performance, cross-platform applications.

Looking back at the history of Swift, it illustrates Apple's dedication to innovation and collaboration. These milestones have not only reshaped the programming landscape but have also set a new precedent for how technology companies engage with their developer communities.

Let's explore the history of the Swift language in more detail to see how it evolved at each step. This will give us a better understanding of how we arrived at Swift 6.

Version	Release date	Description	Key features
Swift 1.0	September 9, 2014	Swift 1.0 was a modern, powerful, and intuitive language designed to make coding faster, safer, and more interactive. It was a major step forward from Objective-C.	• Clean and concise syntax • Type inference • Optionals for safer code • Closures • Automatic memory management • Interactive playgrounds for experimentation and learning
Swift 2.0	September 21, 2015	This release focused on enhancing the language's capabilities and improving developer productivity. Swift 2.2, which was released in December 2015, was also the first open-source version of Swift and the first version to support Linux development with Swift.	• Error handling with try-catch mechanisms • Protocol extensions for adding functionality to types retroactively • Availability checking for platform-specific features • Syntax refinements, including the introduction of guard statements
Swift 3.0	September 13, 2016	Swift 3.0 marked a significant milestone in the language's evolution. This release focused on source compatibility, with the goal of unifying the Swift ecosystem and streamlining the development process.	• Swift Package Manager for managing dependencies and building packages • API design guidelines for consistent code style • Foundation improvements for greater consistency and clarity • The Swift evolution process was defined for community-driven language enhancements • Removal of C-style for loops

Swift 4.0	September 19, 2017	With Swift 4.0, Apple continued its focus on stability, performance, and compatibility. This release introduced several features focused on improving developer productivity and boosting the performance of Swift code.	• Codable protocol for encoding and decoding data • Key paths for type-safe access to properties • Multi-line string literals for improved readability • Enhancements to the String API, including Unicode improvements
Swift 5.0	March 25, 2019	Swift 5.0 represented an important milestone for the language, achieving ABI stability. This release focused on improving the language's performance, reliability, and interoperability. While the previous releases lasted about a year, the 5.x version of Swift has been around for 5 years with 10 "minor" revisions.	• ABI stability for improved compatibility and performance • Result type for handling function results and errors • Raw strings for easier handling of escape characters
	September 20, 2019	Swift 5.1	• Property wrappers for simplifying property access and modification • Implicit returns from single-expression functions • Opaque result types for abstracting implementation details

Swift 5.0	March 24, 2020	Swift 5.2	• Improved diagnostics and error messages • Compiler performance improvements • Enhancements to the Swift standard library
	September 16, 2020	Swift 5.3	• Multiple trailing closures for cleaner syntax • Enum cases with associated values can now be used as standalone types • Enhanced implicit member syntax for more concise code • Support for Windows was officially added to Swift
	March 25, 2021	Swift 5.4	• Enhanced control over implicit member expressions • Enhanced implicit member syntax for more concise code • Improved performance and stability
	September 20, 2021	Swift 5.5	• Async/await for asynchronous programming • Structured concurrency for managing asynchronous tasks • Continuations for managing asynchronous code • Actor isolation for managing concurrent access to mutable state
	March 14, 2022	Swift 5.6	• Introduced existential any type placeholders • Added an Unavailability condition

Swift 5.0	September 12, 2022	Swift 5.7	• Added if/let shorthand for unwrapping optionals • Multi-statement closure type inference • Regular expressions • Supports concurrency in top-level code
	March 30, 2023	Swift 5.8	• Relaxed restrictions on variables in results builders • Opened existential arguments to optional parameters • Concise magic filename • Allows implicit self for weak self captures after self is unwrapped
	September 18, 2023	Swift 5.9	• Use if and switch as expressions • Macros were introduced • Non-copyable structs and enums
	March 5, 2024	Swift 5.10	• Data race safety • Deprecated @UIApplicationMain and @NSApplicationMain • Allows protocols to be nested in a non-generic context • Strict concurrency for global variables

Table 1.1: History of Swift

What's new with Swift 6?

WWDC 2024 marked a significant milestone for the Swift language. Not only was Swift 6 announced, but it also celebrated the 10th anniversary of Apple announcing Swift at WWDC 2014. Swift 6.0 was released in September 2024 and promised to shape the future of the language, building upon the foundation of the previous releases and introducing several new features and enhancements focused on improving developer productivity, performance, and reliability.

One of the major new features of Swift 6.0 is **cross-compilation**, which enables developers to compile code on macOS devices and then deploy it on the Linux platform.

Another important enhancement is the **Swift Testing framework**, which takes advantage of modern Swift features to provide a suite of tools for writing and executing tests, ensuring code quality and reliability.

Additionally, Swift 6.0 introduces **typed throws**, a feature that enhances error handling by enabling functions to specify the types of errors they can throw. This improvement makes error management more explicit and predictable, which will improve code clarity and safety.

Data race safety is another crucial feature, providing built-in mechanisms to prevent concurrent data access issues, which improves the stability and reliability of multi-threaded applications. Strict concurrency checking was also introduced with Swift 6, which enforces rules at compile time to ensure that code adheres to Swift's structured concurrency model, helping developers catch potential issues early and write safer, more predictable concurrent code.

Finally, embedded Swift **expands the language's capabilities into the area of embedded systems**, enabling developers to write Swift code for a much wider range of hardware.

These features, combined with additional enhancements in Swift 6.0, position it as a powerful and versatile language set to drive innovation in software development for years to come.

After Swift 6.0 was announced, Swift continued to improve with the release of Swift 6.1 and Swift 6.2, both bringing powerful new capabilities to the language and its new testing infrastructure. Swift 6.1 extended key paths to support static properties of types. On the testing side, Swift 6.1 added range-based count confirmations, enabling tests to validate on a range in a clear and readable syntax.

With WWDC 2025, and the announcement of Swift 6.2, came several highly requested features, including the addition of the new Observation struct, which adds a flexible way to watch changes in your data outside of SwiftUI. Another important enhancement was the ability to name tasks.

However, one of the most impactful changes was **SE-0466**, which enabled default single-actor execution to effectively set your applications back to being single-threaded. Swift Testing was also expanded with the addition of exit tests using `#expect(processExitsWith:)`, allowing developers to safely test code paths that may terminate the process.

If you have existing projects, it is pretty straightforward to migrate them to Swift 6. Let's take a look at how to do this.

Migrating existing projects to Swift 6

To migrate an existing project to use Swift 6, you generally just need to adjust the building settings for that project. This is how to do it:

1. Open the project in Xcode 16.

2. In the **Project Navigator**, click on the top-level project file.

3. At the top of the editor panel, you should see several tabs: **Info**, **Build Settings**, **Package Dependencies**, and so forth. Click on **Build Settings**.

4. There are a lot of settings within **Build Settings**; we need to scroll down to the **Swift Compiler** settings. To make it easier, we could filter for Swift using the search bar.

5. Under the **Swift Compiler** settings, look for the **Swift Language Version** setting and set it to **Swift 6**.

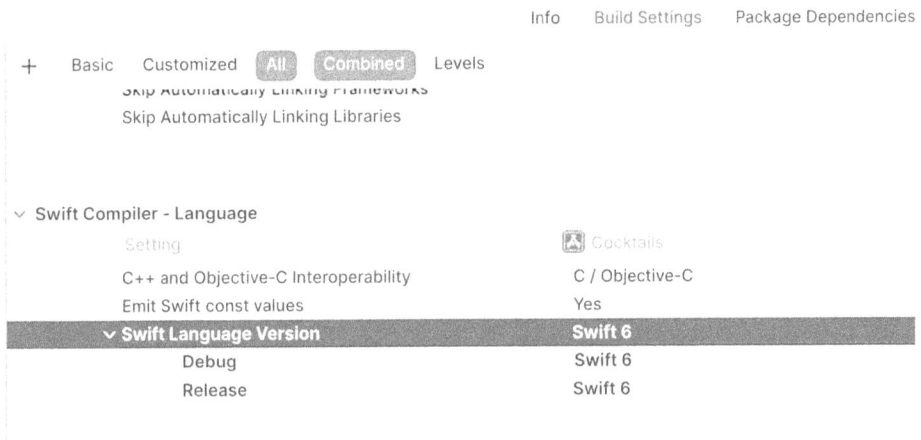

Figure 1.4: Setting the Swift language version to Swift 6

6. Clean and rebuild your project.

That is it; we are now using the Swift 6 compiler to build our project.

Now that we've explored the history of the language leading up to Swift 6.0 and seen how to migrate our project to Swift 6, let's look into ways developers can engage with the Swift community.

How swift.org can help you

When Apple first open-sourced the Swift programming language in 2015, they also launched the swift.org website. This site serves as the central hub for all things related to Swift, providing a variety of resources to support developers at different stages of their journey. Over the years, this site has evolved into a comprehensive resource center, serving as a gateway for developers to explore, learn, and contribute to the Swift ecosystem.

Comprehensive documentation and guides

The swift.org site hosts official documentation along with numerous guides designed to assist developers in exploring the Swift ecosystem. Additionally, it offers detailed articles covering a wide array of Swift topics, including server-side development, compatibility with C++, API design guide, and distinctions between value and reference types. These valuable resources give developers a thorough understanding of the language's features and complexities.

The Swift standard library and Core Libraries

One essential resource within the swift.org site is the documentation for the Swift standard library and the Core Libraries.

These libraries provide a wide range of functions that are usable across all platforms supported by Swift. The standard library contains the essential building blocks for writing Swift programs, while the Core Libraries offer more advanced features.

The Swift Package Manager

The Swift Package Manager is another critical component of the swift.org ecosystem. It simplifies the process of managing and distributing Swift code packages. It also automates the tasks of downloading, compiling, and linking dependencies. With this tool, developers can easily add third-party libraries and frameworks to their projects, making development and the compiling process smoother.

Continuous integration and source compatibility

To keep Swift stable, the swift.org site also manages the project's **continuous integration (CI)** system and source compatibility test suite. These tools are essential for maintaining the language's integrity by conducting thorough tests on active branches and a collection of Swift source code.

Contributing to Swift

For developers who wish to take an active role in shaping the future of Swift, the swift.org site provides information on how to contribute to the community. It explains the Swift evolution process, which decides how Swift evolves, and provides tips on suggesting changes and joining discussions. Additionally, the website shares the source code for the Swift project itself, letting developers look into how Swift works, report problems, and propose improvements with pull requests.

A thriving community

At the core of the swift.org ecosystem is a thriving community of developers, contributors, and enthusiasts. Through forums and communication channels on the website, members can collaborate, help each other, and discuss the future of the Swift language. By fostering this community, the swift.org site has become the center of innovation, helping developers learn, progress, and influence the future of the Swift language.

The swift.org site is a comprehensive resource, offering developers a vast collection of documentation, guides, libraries, and tools for the Swift language. Whether you're just starting out or are already an experienced Swift developer, the site probably has something valuable to offer you, making it a key destination for anyone interested in developing professionally with Swift. I truly hope you explore it and become a part of the Swift community.

Summary

In this chapter, we started by looking at how the Swift language has progressed over time. By understanding the history of the language, we gain a better appreciation of the hard work and dedication that have gone into making the language what it is today. We also get a better understanding of why certain features or design choices were made.

We concluded this chapter with a section introducing the swift.org website, which is an essential resource for the Swift community, offering incredibly valuable documentation, collaboration opportunities, and support for developers of all levels.

In the next chapter, we will explore **closures**, self-contained blocks of functionality, and **result builders**, which allow us to create custom domain-specific languages.

Unlock this book's exclusive benefits now

Scan this QR code or go to packtpub.com/unlock, then
search this book by name.

Note: Keep your purchase invoice ready before you start.

2

Closures and Result Builders

Closures play a critical role in Swift's flexibility and expressiveness. These self-contained blocks of functionality can capture and store references to variables and constants from their surrounding context, making them perfect for implementing callbacks, event handlers, and functional programming patterns. Closures allow us to write concise and readable code that handles complex tasks, such as asynchronous operations and custom sorting algorithms.

Result builders are a more recent addition to Swift that enable us to create custom **Domain-Specific Languages (DSLs)** for constructing complex data structures. With result builders, we can write more expressive and concise code. They do this by allowing us to transform complicated code into clear, readable expressions that look almost like natural language. This not only improves the readability of your code but also makes it easier to organize and maintain complex data structures.

In this chapter, we will explore the mechanics and applications of closures and result builders, with the aim of providing you with a comprehensive understanding of their usage and benefits, from basic syntax and usage patterns to advanced techniques.

In this chapter, we will learn about the following:

- What are closures?
- How to create a closure
- How to use a closure
- What are some examples of useful closures?
- How to use result builders

Let's start by taking a look at closures.

Introducing closures

Closures are self-contained blocks of code that can be passed around and used throughout our application. Like how the **Int** type holds an integer and the **String** type holds a string, a closure can be viewed as a type that holds a block of code. We can, therefore, assign closures to variables, pass them as arguments to functions, and return them from functions.

One of the key features of closures is that they can capture and retain references to any variable or constant from the context in which they were created, a process known as closing over variables or constants. Generally, Swift handles the memory management for us, except in cases where a strong reference cycle is created. We will show you how to resolve strong reference cycles in *Chapter 15, Memory Management*.

Closures in Swift are similar to blocks in Objective-C but are simpler to use and understand. The syntax for defining a closure in Swift is as follows:

```
{
    (<#parameters#>) -> <#return-type#> in <#statements#>
}
```

This syntax resembles how functions are created and, actually, in Swift, global and nested functions are closures. The primary difference between closures and functions is the use of the in keyword. The in keyword replaces curly brackets to separate the closure's parameter and return type definitions from its body. Another difference is a function always has a name while a closure does not.

> 💡 **Quick tip**: Enhance your coding experience with the **AI Code Explainer** and **Quick Copy** features. Open this book in the next-gen Packt Reader. Click the **Copy** button
>
> (**1**) to quickly copy code into your coding environment, or click the **Explain** button
>
> (**2**) to get the AI assistant to explain a block of code to you.

```
                                                          Copy     Explain
function calculate(a, b) {                                 1         2
  return {sum: a + b};
};
```

> 📖 **The next-gen Packt Reader** is included for free with the purchase of this book. Scan the QR code OR go to packtpub.com/unlock, then use the search bar to find this book by name. Double-check the edition shown to make sure you get the right one.

Closures have many applications, which we will explore later in this chapter, but first, let's grasp the basics of closures by examining some basic examples to help us understand what they are, how to define them, and how to use them.

Simple closures

We will start by creating a very simple closure that neither accepts any arguments nor returns any value. Its only task is to print Hello World to the console. The following code shows how to do this:

```
let clos1 = { () -> Void in
    print("Hello World")
}
```

In this example, we define a closure and assign it to the clos1 constant. Since no parameters are specified within the parentheses, this closure does not accept any arguments. Additionally, its return type is Void, indicating that it does not return any value. The closure's body consists of a single line that prints Hello World to the console.

Closures can be used in many ways; in this example, we simply want to execute it. To do so, we can call the closure like this:

```
clos1()
```

After executing the closure, Hello World will be printed to the console. While closures may not seem particularly useful at this stage, we will discover their versatility and power as we progress through this chapter.

Let's look at another simple example. This next closure will accept a single parameter of the String type named name but will not return any value. Inside the closure, we will print a greeting to the name passed into the closure. Here is the code for this second closure:

```
let clos2 = {
    (name: String) -> Void in
    print("Hello \(name)")
}
```

The main difference between this closure and the previous example is that this closure defines a single parameter of the String type within the parentheses. Parameters for closures are defined similarly to parameters of functions.

We can execute this closure just as we executed the previous one. The following code demonstrates this:

```
clos2("Jon")
```

When the closure is executed, it will print Hello Jon to the console.

> An important point to note is that named parameters in a closure do not require the parameter name to be used, and in fact, you will get an error if you use the name when the closure is called.

Our original definition of closures stated that they are self-contained blocks of code that can be passed around and used throughout our application. This implies that we can pass closures from the context in which they were created into other parts of our code. Let's see how to pass our clos2 closure into a function. We will define a function that accepts our closure as follows:

```
func testClosure(handler: (String) -> Void) {
    handler("Luna")
}
```

We define this function like any other, but in the parameter list, we include a parameter named handler of the (String) -> Void type. This matches the parameter and return types we defined for the clos2 closure, enabling us to pass this closure into the function. Here's how we could do this:

```
testClosure(handler: clos2)
```

We call the testClosure() function as we would other functions, and the closure being passed in is treated like any other variable. Since the clos2 closure is executed within the testClosure() function, the message Hello Luna will be printed to the console when this code is executed. As we will see later in this chapter, the ability to pass closures to functions is what makes them so powerful and versatile.

Finally, let's look at how to return a value from a closure. The following example shows this:

```
let clos3 = {
    (name: String) -> String in
    return "Hello \(name)"
}
```

The clos3 closure definition is like the previous one, except that the return type is changed from Void to String. Instead of printing the message to the console, the closure returns the message using the return statement. We can execute this closure just like the previous closures, or pass it to a function. Here's an example of executing the closure:

```
var message = clos3("Maple")
print(message)
```

After this line of code is executed, the message variable will contain the string `Hello Maple`. These examples illustrate the format and definition of closures. While the closure syntax we've shown so far is concise, it can be made even shorter and simpler. In the next section, we will look at the shorthand syntax for closures.

Shorthand syntax for closures

The use of shorthand syntax for closures is largely a matter of personal preference. Many developers enjoy making their code as compact as possible and take great pride in doing so. However, this can sometimes make the code harder for other developers to read and understand.

One of the most popular shorthand syntaxes for closures is the one that is primarily used for sending small, usually one line, closures to a function. Before we explore this shorthand syntax, let's write a function that accepts a closure as a parameter:

```
func testFunction(num: Int, handler: () -> Void) {
    for _ in 0..<num {
        handler()
    }
}
```

This function accepts two parameters: an integer named `num` and a closure named `handler`, which takes no parameters and returns no value. Inside the function, we use a `for` loop that iterates `num` times, and within this loop, the closure that was passed into the function is called. Now, let's create a closure and pass it to `testFunction()`, as follows:

```
let clos = { () -> Void in
    print("Hello from standard syntax")
}
testFunction(num: 5, handler: clos)
```

This code is clear and easy to understand; however, it spans five lines. Now, let's see how we can shorten it by writing the closure inline within the function call:

```
testFunction(num: 5, handler: { print("Hello from shorthand closure") })
```

In this example, we created the closure inline within the function call. When we create a closure inline like this, the closure is enclosed in curly brackets ({}), making the code to create this closure `{ print("Hello from shorthand closure") }`. When executed, this code will print `Hello from shorthand closure` five times.

The optimal way to call testFunction() with a closure, balancing both compactness and readability, could be something like this:

```
testFunction(num: 5) {
    print("Hello from Shorthand closure")
}
```

Having the closure as the final parameter enables us to omit the label when calling the function, making the code both compact and readable. This shorthand is called a trailing closure.

Now, let's see how to use parameters with this shorthand syntax. Let's start by creating a new function named testFunction2 that accepts a closure with a single parameter. The following example shows how we may do this:

```
func testFunction2(num: Int, handler: (_: String) -> Void) {
    for _ in 0..<num {
        handler("Me")
    }
}
```

In testFunction2, we define the closure as (_: String) -> Void. This definition means that the closure accepts one parameter and does not return any value.

Now, let's see how we could use this shorthand syntax to call this function:

```
testFunction2(num: 5) {  name in
    print("Hello from \(name)")
}
```

This code calls the testFunction2() function with a value of 5 for the first argument and a closure for the second argument. This closure takes a parameter named name.

Now, let's see how we can make it even more compact:

```
testFunction2(num: 5) {
    print("Hello from \($0)")
}
```

The key difference between this closure and the previous one is the use of $0. The $0 parameter is shorthand for the first parameter that is passed into the closure. When this code is executed, it prints Hello from Me five times. Using the dollar sign ($) followed by a number allows us to reference parameters without explicitly creating a parameter list in the definition. The number after the dollar sign indicates the position of the parameter in the parameter list.

Let's look at this format further, as we are not limited to using the dollar sign with a number shorthand format only with inline closures. This shorthand syntax can also simplify the closure definition by omitting parameter names. The following example shows how this may work:

```
let clos5: (String, String) -> Void = {
    print("\($0) \($1)")


}
```

In this example, the closure has two parameters of the String type defined; however, they are not named. The parameters are specified as (String, String). Within the body of the closure, we can access these parameters using $0 and $1. Notice that the closure definition appears after the colon (:), similar to how we define a variable type, rather than inside the curly brackets. This is the proper way to define a closure when using anonymous arguments because with type inference, a closure knows the number and type of arguments. Defining the closure with anonymous arguments in any other way, such as the following, would not be valid:

```
let invalidClosure = { (String, String) in
    print("\($0) \($1)")
}
```

If we tried to run this code, we would receive an error letting us know that this is not valid.

Next, let's look at how we would use the clos5 closure:

```
clos5("Hello", "Kai")
```

Since Hello is the first string in the parameter list, it is accessed using $0, and Kai, being the second string, is accessed using $1. When this code is executed, the message Hello Kai will be printed to the console. In the next example, when the closure does not return any value, we can use parentheses instead of defining the return type as Void:

```
let clos6: () -> () = {
    print("Howdy")
}
```

In this example, we define the closure as () -> (). This indicates that the closure neither accepts any parameters nor returns a value. We can execute this closure as follows:

```
clos6()
```

Personally, I am not very fond of this shorthand syntax. I find the code much easier to read when the Void keyword is used instead of parentheses.

Before we look at some useful examples of closures, there's one more shorthand example to cover. In this example, we'll demonstrate how to return a value from a closure without needing to use the return keyword. If the entire closure body consists of only a single statement, we can omit the return keyword, just as we do with functions, and the result of the statement will be returned automatically. The following example shows how we may do this:

```
let clos7 = { (first: Int, second: Int) -> Int in first + second }
print(clos7(1,2))
```

In this example, the closure accepts two parameters of the integer type and returns an integer. The closure's body consists of a single statement that adds the two integers together. Notice that we do not include the return keyword before the addition statement. The compiler recognizes that this is a single-statement closure and will automatically return the result, just as if we had included the return keyword. The result type of our statement must match the closure's return type in order to be valid.

The examples shown in the previous sections were designed to demonstrate how to define and use closures. On their own, these examples do not fully showcase the power and usefulness of closures. Let's look at some examples that will highlight the power and practicality of closures in Swift.

Using closures with Swift arrays

Swift arrays come with several built-in algorithms that use closures. Now that we have a good understanding of how closures work, let's see how we can use closures to make full use of these algorithms.

In this section, we will primarily use the map algorithm for consistency; however, the concepts we demonstrate can be applied to any of the algorithms.

Let's begin by defining an array to work with:

```
let guests = ["Jon", "Heidi", "Kailey", "Kai"]
```

This array, named guests, contains a list of names and will be used for most of the examples in this section.

Now that we have our example array, let's add a closure that will print a greeting for each name in the array:

```
guests.map { name in
    print("Hello \(name)")
}
```

Where the map algorithm applies the closure to each item in the array, this example will print the greeting for each name in the array. After the first section of this chapter, you should have a solid understanding of how this closure works. Using the shorthand syntax we discussed earlier, we can simplify the previous example to a single line of code:

```
guests.map {print("Hello \($0)")}
```

This is an example where the shorthand syntax, in my opinion, is easier to read than the full syntax.

Now, let's say that instead of printing the greeting to the console, we want to return a new array with the greetings. To do this, we could modify the closure to return a String type, as shown in the following example:

```
var messages = guests.map {
    (name:String) -> String in
    return "Welcome \(name)"
}
```

When this code is run, the messages array will contain a greeting for each name in the guests array, while the original guests array will remain unchanged.

In the previous examples, we demonstrated how to add a closure to the map algorithm inline. This approach works well if there is only one closure to use with the algorithm; however, if we have multiple closures, we need to use the same closure multiple times, or if we reuse the closure with different arrays, we can assign the closure to a constant. This allows us to reference the closure by its constant name where needed.

Let's see how we could do this by defining two closures. One closure will print a greeting for each element in the array, while the other will print a goodbye message for each element:

```
let greetGuest = { (name:String) -> Void in
    print("Hello guest named \(name)")
}
let sayGoodbye = { (name:String) -> Void in
    print("Goodbye \(name)")
}
```

Now that we have defined two closures, we can use them with the map algorithm as needed. The following code demonstrates how to use these closures interchangeably with the guests array:

```
guests.map(greetGuest)
guests.map(sayGoodbye)
```

When we use the greetGuest closure with the guests array, it prints the greeting message to the console, and when we use the sayGoodbye closure with the guests array, it prints the goodbye message to the console. If we had another array named guests2, we could use the same closures with that array as well.

All the examples in this section so far have been to either print a message to the console or return a new array from the closure. However, we are not limited to such basic functionality. For instance, we can filter the array within the closure, as demonstrated in the following example:

```
let greetGuest2 = {
    (name:String) -> Void in
    if (name.hasPrefix("K")) {
        print("\(name) is on the guest list")
    } else {
        print("\(name) was not invited")
    }
}
```

In this example, we print a different message depending on whether the name starts with the letter "K" or not.\

As mentioned earlier in the chapter, closures can capture and store references to any variable or constant from the context in which they are defined. Let's look at an example of this. Suppose we have a function that contains the highest temperatures for the past seven days at a given location and accepts a closure as a parameter. This closure will do some sort of analysis of the temperatures. To start with, the analyzeTemperature() function will look like this:

```
func analyzeTemperatures(analysis:([Int])->Void) {
    let tempArray = [72,74,76,68,70,72,66]
    analysis(tempArray)
}
```

This function includes a list of high temperatures and processes them using the closure passed into the function. Now, let's create the closure that will analyze the temperatures and execute the function:

```
let threshold = 71

let daysAboveTemperature = { (temperatures: [Int]) in
    var aboveThresholdCount = 0
    for temperature in temperatures {
```

```
        if temperature > threshold {
            aboveThresholdCount += 1
        }
    }
    print("Number of days above \(threshold)°F: \(aboveThresholdCount)")
}
```

In this code, notice that the threshold constant is defined outside the closure but is used within it. The closure captures this constant from its surrounding context and uses it to count how many days the temperatures exceed this threshold. This demonstrates the closure's ability to reference and use variables from the context in which it was defined.

While closures have the ability to update variables in the context within which they were created, they cannot modify items within the function's context. This means it cannot alter variables within the analyzeTemperatures function. However, we could move the aboveThresholdCount variable outside of the closure, as shown in this code:

```
let threshold = 71
var aboveThresholdCount = 0

let daysAboveTemperature = { (temperatures: [Int]) in
    for temperature in temperatures {
        if temperature > threshold {
            aboveThresholdCount += 1
        }
    }
    print("Number of days above \(threshold)°F: \(aboveThresholdCount)")
}

analyzeTemperatures(analysis: daysAboveTemperature)
```

The output of this code is the same as the previous one with the aboveThresholdCount variable within the closure.

We mentioned trailing closures earlier in this chapter. Another powerful feature of closures is the ability to have multiple trailing closures. Let's look at this.

Multiple trailing closures

A function can accept multiple closure parameters. When these closures are used as trailing clo-sures, they can be written outside of the function's parentheses, as we saw with a single trailing closure. Each trailing closure is labeled in order to define which parameter it corresponds to. This is especially useful in scenarios where a function may handle different outcomes or behaviors, such as success and failure. Let's look at an example of this:

```
func performTask(success: () -> Void, failure: () -> Void) {
    let taskSucceeded = Bool.random()

    if taskSucceeded {
        success()
    } else {
        failure()
    }
}
```

In this example, the performTask() function accepts two closures as parameters, one named success and one named failure, and then within the function, depending on if the logic suc-ceeds or fails, the appropriate closure is run. In this example, we use a random value to determine success or failure. We could then use this function like this:

```
performTask {
    print("Task succeeded!")
} failure: {
    print("Task failed!")
}
```

With this code, if performTask succeeds, the Task succeeded! method is printed out; otherwise, the Task failed! message is shown.

Now that we have seen how to use multiple trailing closures, let's look at an advanced use case for closures.

Advanced closures example

There are numerous use cases where closures can be used to create flexible and extensible code, such as handling asynchronous tasks, managing callbacks, and implementing event-driven programming. In this section, we will show a couple of advanced techniques using closures, while demonstrating how closures can be used to create a basic logging system that handles different levels of log messages.

Let's start by creating an enumeration that will define the log levels and a Logger class:

```swift
enum LogLevel {
    case info, warning, error
}

class Logger {
    typealias logLevelHandler = (String) -> Void

    private var handlers: [LogLevel: [logLevelHandler]] = [:]

    func registerHandler(for level: LogLevel,
                         handler: @escaping logLevelHandler) {
        if handlers[level] == nil {
            handlers[level] = []
        }
        handlers[level]?.append(handler)
    }

    func log(_ message: String, level: LogLevel) {
        if let levelHandlers = handlers[level] {
            for handler in levelHandlers {
                handler(message)
            }
        }
    }
}
```

This code defines a basic Logger type that allows us to register a log handler and also to log a message. The enumeration defines three log levels.

There are a couple of items to note about this code. First, a typealias is created for the closure definition. This enables us to use the alias anywhere we need to define or pass the closure, making the code easier to read and maintain.

The second item worth pointing out is the use of the @escaping keyword in the registerHandler() function. This keyword is used to indicate that the closure being passed into the function may be called after the function itself has returned. This is important for closures that might be stored or executed asynchronously, such as those used in completion handlers, timers, or background tasks.

Now, let's look at how we can add log handlers to the logging class. The following code demonstrates how to create three very simple handlers:

```swift
let logger = Logger()

logger.registerHandler(for: .info) { message in
    print("INFO: \(message)")
}
logger.registerHandler(for: .warning) { message in
    print("WARNING: \(message)")
}

logger.registerHandler(for: .error) { message in
    print("ERROR: \(message)")
}
```

In this code, we create an instance of the Logger type and register three log handlers. The first is set to .info, the second to .warning, and the third to .error. Each handler simply prints a message to the console.

If we needed more complex log handlers, we could create them and add them to the Logger class like this:

```swift
let emailError = {
    (message: String)  -> Void in

    //code to email error
    print("Emailed Error: \(message)")
}
logger.registerHandler(for: .error, handler: emailError)
```

Here, we create a more complex log handler by defining it outside of the registerHandler() function call. This allows us to create more complex handlers than we would create inline and also enables us to reuse the log handler in other parts of our code. However, as we can see, this log handler also simply prints a message to the console.

Finally, now that the log handlers have been added, we can begin to log our messages, as shown here:

```
logger.log("Informational Message", level: .info)
logger.log("Warning message", level: .warning)
logger.log("We have an error", level: .error)
```

This code will display the log messages to the console as well as email the error message if the email functionality is implemented.

Closures are a powerful and flexible feature that allows us to encapsulate functionality in a self-contained block of code and pass it around within our application. By understanding and utilizing closures effectively, we can improve our ability to create modular, reusable, and maintainable code. Whether we are using closures as callback functions, event handling, or within the functional programming patterns, mastering closures is essential for any Swift developer who wants to utilize the full potential of the language. Now, let's look at result builders, which is a feature that was added in Swift 5.4.

Result builders

Result builders allow developers to create custom DSLs for constructing complex data structures, such as JSON, SwiftUI views, or HTML elements for server-side Swift frameworks. They help us write code that is more expressive and concise, making it easier to understand and maintain.

To get a better understanding of what result builders can do, let's look at a real basic example, one that will combine multiple strings:

```
@resultBuilder
struct StringBuilder {
    static func buildBlock(_ components: String...) -> String {
        return components.joined()
    }
}

func buildString(@StringBuilder _ components: () -> String) -> String {
    return components()
}
```

In this example, we use the `@resultBuilder` attribute to define a custom result builder. In this case, it's applied to the `StringBuilder` structure, indicating that this structure is intended to be used as a result builder.

Within the `StringBuilder` structure, we define the static `buildBlock()` method, which is a special method that is recognized by result builders. For this example, the `buildBlock()` method takes a variadic parameter of type `String` (indicated by `String...`) and returns an instance of the `String` type. This method is responsible for concatenating all the strings passed as parameters using the `joined()` method.

The `buildString()` function takes a closure as a parameter, marked with the `@StringBuilder` attribute. This means that the closure can be processed by the `StringBuilder` result builder. The closure returns an instance of the String type. The function itself returns the result of that closure's execution.

This string builder could be used as shown in the following example:

```
let result = buildString {
    "Hello, "
    "Mastering "
    "Swift!"
}


print("\(result)")
```

This code calls the `buildString()` function with a closure as its argument. The content of the closure is three string literals, which are separated by line breaks but are not explicitly concatenated with any operator. The last line prints the results, which look like this:

```
Hello, Mastering Swift!
```

Now that we have seen how to define a simple result builder, let's look at a more complex example, one that will take JSON data and put it into a dictionary. We will start with the following code, which defines our result builder:

```
struct DictionaryComponent {
    let dictionary: [String: Any]

    func addToJSON(_ json: inout [String: Any]) {
        for (key, value) in dictionary {
            json[key] = value
```

```
                }
            }
        }


@resultBuilder
struct JSONBuilder {
    static func buildBlock(_ components: DictionaryComponent...) ->
                            [String: Any] {
        var result: [String: Any] = [:]
        for component in components {
            component.addToJSON(&result)
        }
        return result
    }


    static func buildExpression(_ expression: [String: Any]) ->
                                DictionaryComponent {
        return DictionaryComponent(dictionary: expression)
    }
}
```

This code begins by creating a structure named `DictionaryComponent`, which is used to add key-value pairs to a JSON dictionary. The `addToJSON()` method is used to add these key-value pairs from its dictionary property to a `json` dictionary parameter provided as an `inout` parameter.

The `JSONBuilder` structure includes the `buildBlock()` method, which is similar to the one provided in the `StringBuilder` structure shown in the previous example. However, this `buildBlock()` method is a little different because it iterates over each `DictionaryComponent` in the `components` parameter and invokes the `addToJSON()` method on each component to add its contents to the **result** dictionary.

There is an additional method within the `JSONBuilder` structure that was not in the `StringBuilder` structure, which is `buildExpression()`. This method takes a single parameter of type `[String: Any]`, which represents a dictionary and returns a `DictionaryComponent`. It is responsible for handling individual dictionary expressions, by creating a new `DictionaryComponent` using the provided dictionary and returning it as a `DictionaryComponent`.

In Swift's result builder mechanism, there are two types of expressions that can be processed:

- Block expressions: These are expressions contained within a block, such as multiple components passed to a function. The buildBlock() method in the result builder is responsible for handling these block expressions.
- Individual expressions: These are expressions that stand alone, outside of a block. The buildExpression() method in the result builder is responsible for handling these individual expressions.

In the case of the JSONBuilder structure in the previous example:

- The buildBlock() method is used to handle block expressions, where multiple DictionaryComponent instances are passed as variadic parameters.
- The buildExpression() method is used to handle individual expressions, where a single dictionary is passed as an argument.

The buildExpression() method in the JSONBuilder structure allows it to handle both individual and block expressions uniformly within the result builder context. This allows for more flexibility when constructing JSON-like structures, as we can use either individual expressions or blocks of expressions interchangeably.

If our JSON data contains nested dictionaries, the buildExpression() method assists in handling these nested components. For instance, if your input dictionary contains nested dictionaries as values, the buildExpression() method can recursively handle each nested dictionary, wrapping them into DictionaryComponent instances and adding them to the final JSON dictionary using the addToJSON() method.

We would use JSONBuilder like this:

```
@JSONBuilder
func buildJSON() -> [String: Any] {
    [
        "name": "Jon",
        "age": 30,
        "address": [
            "city": "Boston",
            "zipcode": "10001"
        ]
    ]
```

```
    }


    let json = buildJSON()
    print(json)
```

The JSONBuilder annotation is used with the buildJSON() function to indicate that it should be processed using the JSONBuilder result builder. This function constructs and returns a dictionary object that is built from the JSON data defined within the function.

When the buildJSON() function is called, the result builder processes the dictionary components provided by the function, and the results are stored within the json constant, which is then printed to the console. The results should look like this:

```
["name": "Jon", "address": ["city": "Boston", "zipcode": "10001"], "age": 30]
```

Note, the keys are not ordered therefore the results could appear in different orders when the example is run.

Swift's result builders enable us to construct complex data structures clearly and concisely. With the ability to define custom result builders, Swift provides flexibility for constructing various types of data, such as strings, JSON, HTML, and even custom DSLs. By using result builders effectively, we can write cleaner, more modular, and more maintainable code.

Summary

In this chapter, we provided an in-depth introduction to closures, which are self-contained blocks of code that can be passed around within our applications. We saw how closures can capture and retain references to variables or constants from the context in which they are created. Starting with simple examples, we learned how to define and execute closures, including ones that accept parameters and return values. We also touched on how closures can be used to increase code flexibility by passing them as function arguments and reusing them across different contexts, and we created a basic Logger type that demonstrated the power and flexibility of closures.

We also introduced result builders, a feature that allows us to create custom DSLs used for defining complex data structures. We saw how result builders, like closures, enhance code expressiveness and flexibility, making it easier to work with data structures such as JSON, HTML, and SwiftUI views. The examples in this section demonstrated how result builders can simplify complex tasks, emphasizing their role in writing cleaner and more modular code.

In the next chapter, we will look at how protocols and protocol extensions play a crucial role in writing flexible and easy-to-maintain code.

Unlock this book's exclusive benefits now

Scan this QR code or go to `packtpub.com/unlock`, then search this book by name.

Note: Keep your purchase invoice ready before you start.

3

Protocols and Protocol Extensions

In Swift, protocols and protocol extensions play crucial roles in the writing of code that is flexible, reusable, and easy to maintain. Protocols act as blueprints, defining methods, properties, and other requirements that a type must adhere to when conforming to the protocol. This allows for the creation of standardized interfaces, which enables seamless communication between different components of an application and promotes code interoperability.

Additionally, protocol extensions provide a mechanism to extend the functionality of types without altering the original implementation and provide default implementations for methods and computed properties defined in a protocol. This enables developers to enhance existing types, including those from third-party libraries or Swift's built-in types, promoting better code organization and scalability.

In this chapter, you will learn about the following topics:

- How protocols are used as a type
- How to implement polymorphism in Swift using protocols
- How to use protocol extensions
- What existential anys are, and how we use them

In Swift, classes, structures, and enumerations can all conform to protocols; we will be focusing on classes and structures in this chapter. Enumerations should mainly be used when we need to represent a finite set of cases, and while there are valid situations where we would have an enumeration conform to a protocol, they are less common in my experience. It is important to note that anywhere we refer to a class or structure, we could also use enumerations.

To really grasp how useful protocols and protocol extensions are, let's get a better understanding of them. We'll begin by seeing how protocols are full-fledged types in Swift.

Protocols as types

Even though protocols do not include any functionality, they're regarded as full-fledged types in Swift and can be used just like any other type. This means we can use protocols as parameter types or return types in functions. They can also serve as the type for variables, constants, and collections.

Let's look at some examples to illustrate what that means. For these few examples, we will use the following `Person` protocol:

```
protocol Person {
    var firstName: String { get set }
    var lastName: String { get set }
    var birthDate: Date { get set }
    var profession: String { get }

    init(firstName: String,lastName: String, birthDate: Date)
}
```

The `Person` protocol requires conforming types to have four properties: `firstName`, `lastName`, and `birthDate` (all of which must be both readable and writable (`{ get set }`)), and `profession`, which is read-only (`{ get }`). Additionally, the protocol requires types that conform to this protocol to provide an initializer that takes three parameters: `firstName`, `lastName`, and `birthDate`, of the `String` or `Date` type. By defining these requirements, the protocol ensures that any type conforming to it will have a consistent interface, while also ensuring that the type can be properly initialized.

In this first example, the `Person` protocol is used as both a parameter type and a return type for a function:

```
func updatePerson(person: Person) -> Person {
    // Code to update person goes here
    return person
}
```

In this example, the `updatePerson()` function accepts one parameter of the `Person` protocol type and returns a value of this type.

This next example shows how to use a protocol as a type for constants or variables:

```
var myPerson: Person
```

In this example, we create a variable of the Person protocol type, which we name myPerson.

Protocols can also be used as the item type for storing a collection, such as arrays, dictionaries, or sets:

```
var people: [Person] = []
```

In this final example, we created an array of the Person protocol type.

As we can see from these examples, even though the Person protocol does not implement any functionality, we can still use protocols when we need to specify a type. However, a protocol cannot be instantiated in the same way as a class or a structure. The reason for this is that the protocol does not implement any functionality. As an example, when trying to create an instance of the Person protocol, as shown in the following example, we would receive a compile-time error, stating that the type person cannot be instantiated:

```
var test = Person(firstName: "Jon", lastName: "Hoffman", birthDate:Date())
```

We can use the instance of any type that conforms to our protocol wherever the protocol type is required. For example, if a variable is defined to be of the Person protocol type, we can then populate that variable with any class, structure, or enumeration that conforms to this protocol.

To demonstrate this, let's assume that we have two types, named SwiftProgrammer and FootballPlayer, that conform to the Person protocol:

```
var myPerson: Person
myPerson = SwiftProgrammer(firstName: "Jon", lastName: "Hoffman",
    birthDate: Date())
print("\(myPerson.firstName) \(myPerson.lastName)")

myPerson = FootballPlayer(firstName: "Dan", lastName: "Marino",
    birthDate:Date())
print("\(myPerson.firstName) \(myPerson.lastName)")
```

In this example, we begin by creating the myPerson variable of the Person protocol type. We then set the variable to an instance of the SwiftProgrammer type and print the first and last names. Next, we set the myPerson variable to an instance of the FootballPlayer type and also print the first and last names.

Note that Swift does not care whether the instance is a class, structure, or enumeration. The only requirement is that the type conforms to the `Person` protocol type.

The `Person` protocol can also be used as the type for an array, which means that we can populate the array with instances of any type that conforms to the protocol. Once again, the only requirement is that the type conforms to the `Person` protocol.

Now, let's look at how we can use protocols to achieve polymorphism.

Polymorphism with protocols

What we saw in the previous examples is a form of polymorphism. The word **polymorphism** comes from the Greek root **poly**, meaning many, and **morphe**, meaning form. In the programming field, polymorphism is the ability to use a single interface to access multiple types. In the previous examples, the single interface was represented by the `Person` protocol and the multiple types were any type that conforms to that protocol.

Polymorphism gives us the ability to interact with various types with a standard interface. To illustrate this, we can extend the previous example, where we created an array of the `Person` types and looped through it. We can then access each item in the array using the interface defined by the `Person` protocol, regardless of the actual type. Let's see an example of this, assuming we have an array of `Person` types named `people`:

```
for person in people {
    print("\(person.firstName) \(person.lastName):\(person.profession)")
}
```

When we use a protocol as the type of a variable, constant, collection type, and so on, we can use an instance of any type that conforms to that protocol. This is a very important concept to comprehend and is one of the many things that make protocols and protocol extensions so powerful.

When we use the interface defined by a protocol to access instances, as shown in the previous example, we are limited to using only properties and methods that are defined within the protocol itself. If we wanted to use properties or methods that are specific to the individual types, we would need to cast the instance to that specific type.

Typecasting with protocols

Typecasting is a way to check the type of the instance and/or to treat the instance as a specific type. In Swift, we would use the `is` keyword to check whether an instance is a specific type and the `as` keyword to treat an instance that conforms to a protocol as a concrete type.

Let's see how we would check the instance type using the `is` keyword. The following example shows how this is done:

```
for person in people {
    if person is SwiftProgrammer {
        print("\(person.firstName) is a Swift Programmer")
    }
}
```

In this example, we use the `if` conditional statement to check whether each element within the `people` array is an instance of the `SwiftProgrammer` type and, if so, we print that the person is a Swift programmer.

As previously mentioned, the `as` keyword is used to treat the instance as a specific type. In the next example, we use the `if-let` conditional statement to check the instance type:

```
for person in people {
    if let p = person as? SwiftProgrammer {
        print("\(p.firstName) is a Swift Programmer")
    }
}
```

While these examples are good methods to check whether we have an instance of a specific type, they are not very efficient when checking for multiple types. It would be more efficient to use a `switch` statement, as shown in the next example:

```
for person in people {
    switch person {
        case is SwiftProgrammer:
            print("\(person.firstName) is a Swift Programmer")
        case is FootballPlayer:
            print("\(person.firstName) is a Football Player")
        default:
            print("\(person.firstName) is an unknown type")
    }
}
```

In this example, we showed how to use the `switch` statement to check the instance type for each element of the array. For this, we use the `is` keyword in each `case` statement in an attempt to match the instance type.

The `where` statement can also be used with the `is` keyword to filter the array, as shown here:

```
for person in people where person is SwiftProgrammer {
    print("\(person.firstName) is a Swift Programmer")
}
```

Now let's see how we can use protocols with enumerations in Swift.

Using protocols with enumerations

As we mentioned in the introduction to this chapter, we can use protocols with enumerations to enforce consistency across different enumerations. By conforming to a protocol, enumerations can define properties or methods that provide additional functionality while maintaining type safety. This is particularly useful when working with enumerations that need computed properties, default behaviors, or require interaction with other types in a structured way. If you are not familiar with enumerations, we look at them in *Chapter 6, Enumerations*.

Enumerations conform to protocols exactly like structures and classes do by using the colon at the end of the enumeration declaration and then adding the name of the protocol. Let's look at an example of this.

```
protocol Describable {
    var description: String { get }
}


enum VehicleType: Describable {
    case car, bike

    var description: String {
        switch self {
        case .car:
            return "Vehicle with 4 wheels"
        case .bike:
            return "Vehicle with 2 wheels"
        }
    }
}
```

In this example we declare a protocol named `Describable`, which requires a computed property named `description`. We then add the computed property to our `VehicleType` enumeration. We will go more in-depth to enumerations in *Chapter 6, Enumerations*.

A protocol that is created for an enumeration can also be used for other types. For example, we could use the **Describable** protocol with a structure like this.

```
struct Vehicle: Describable {
    var description: String
}
```

We have the ability to use protocols with different types (classes, structures, and enumerations), making protocols very powerful in Swift.

Now that we have seen the basics of protocols, let's look into one of the most exciting features of Swift: protocol extensions.

Protocol extensions

Protocol extensions enable us to extend a protocol in order to provide method and property implementations for conforming types. This enables us to provide common implementations for all the conforming types, eliminating the need to provide separate implementations for each individual type. While protocol extensions might not initially appear exciting, once you see how powerful they can really be, they will transform the way you think about and write code.

Let's begin by looking at how we can use protocol extensions within a very simple example. We will start by defining a protocol named `Dog`, as shown here:

```
protocol Dog {
    var name: String { get set }
    var color: String { get set }
}
```

With this protocol, we state that any type that conforms to the `Dog` protocol must have two properties of the `String` type, named `name` and `color`. Next, let's define the three types that conform to this `Dog` protocol. We will name these types `JackRussel`, `WhiteLab`, and `Mutt`. The following code shows how we would define these types:

```
struct JackRussel: Dog{
    var name: String
    var color: String
```

```
    }
class WhiteLab: Dog{
    var name: String
    var color: String
    init(name: String, color: String) {
        self.name = name
        self.color = color
    }
}
struct Mutt: Dog{
    var name: String
    var color: String
}
```

We purposely created the JackRussel and Mutt types as structures and the WhiteLab type as a class to highlight the differences between how the two types are set up, and to illustrate how they are treated the same when it comes to protocols and protocol extensions.

The biggest difference we can see in this example is that structure types provide a default initiator, whereas with the class we must provide the initiator to populate the properties.

Now, if we wanted to provide a method named speak for each type that conforms to the protocol, prior to protocol extensions, first we would have to add the method definition to the protocol:

```
protocol Dog{
    var name: String { get set }
    var color: String { get set }
    func speak() -> String
}
```

Once the method was defined within the protocol, we would then need to provide an implementation of the method for each type that conformed to the protocol. Depending on the number of confirming types, this could take a lot of coding to implement. The following code shows how we may implement this method for every type:

```
struct JackRussel: Dog{
    var name: String
    var color: String
    func speak() -> String {
```

```
            return "Woof"
        }
    }

    class WhiteLab: Dog{
        var name: String
        var color: String
        init(name: String, color: String) {
            self.name = nameself.color = color
        }
        func speak() -> String {
            return "Woof"
        }
    }

    struct Mutt: Dog{
        var name: String
        var color: String
        func speak() -> String {
            return "Woof Woof"
        }
    }
```

While this works, it is very inefficient because every time we updated the protocol, we would need to update every type that conformed to it, and therefore duplicate a lot of code, as shown in this example. Additionally, if we needed to change the default behavior of the speak() method, we would have to go into each implementation and change the method. Depending on the number of types that conformed to this protocol, this could take a bit of time to implement, and it could affect a lot of code. This is where protocol extensions can assist us.

With protocol extensions, we can remove the speak() method from the protocol itself and define it, with the default behavior, within a protocol extension.

> If a method is implemented within a protocol extension, we are not required to define it within the protocol.

The following code shows how we would define the protocol and the protocol extension:

```
protocol Dog{
    var name: String { get set }
    var color: String { get set }
}


extension Dog{
    func speak() -> String {
      return "Woof Woof"
    }
}
```

In this example, we begin by defining the Dog protocol with the original two properties. We then create a protocol extension that extends the Dog protocol and contains the default implementation of the speak() method. Now, there is no need to provide an implementation of the speak() method within every type that conforms to the Dog protocol because they automatically receive the implementation, from the protocol extensions, as part of the protocol.

Let's see how this works with the following code:

```
let dash = JackRussel(name: "Dash", color: "Brown and White")
let lily = WhiteLab(name: "Lily", color: "White")
let maple = Mutt(name: "Buddy", color: "Brown")

let dSpeak = dash.speak() // returns "woof woof"
let lSpeak = lily.speak() // returns "woof woof"
let bSpeak = maple.speak() // returns "woof woof"
```

With this example, we can see that by adding the speak() method with the Dog protocol extension, every type that conforms to this protocol now receives this functionality. The speak() method in the protocol extension actually acts as the default implementation because we have the ability to override it within the type implementation. As an example, we could override the speak() method in the Mutt structure, as shown here:

```
struct Mutt: Dog{
    var name: String
    var color: String
    func speak() -> String {
```

```
        return "I am hungry"
    }
}
```

When we call the speak() method for an instance of the Mutt type, it will return the I am hungry string.

Now that we have seen how to use protocols and protocol extensions, let's look at Any and any in Swift.

Any and any

Swift uses the Any type in two distinct ways. Notice that one has an uppercase "A" and the other has a lowercase "a." However, these two usages of Any do not have a direct relationship with each other. Let's look at the differences between these two in Swift.

Uppercase "Any"

The uppercase Any in Swift is a special type that can represent an instance of any type, including value types, reference types, and even optional types. What this means is that you can use a variable or constant defined with Any to hold a value of any type, without needing to know the specific type at compile time. When using Any, as the type for an array or a dictionary, we can use different types within the collection. However, when we want to work with these stored values, we need to use type casting to convert them to the specific type. Here's an example of using Any to store values of different types in an array:

```
var mixed: [Any] = [42, "Hello", true]
```

In this example, the mixed array holds an integer, a string, and a boolean. This use of Any has been seen in Swift almost from the very beginning. Now let's look at what the lowercase any does.

Existential "any" in Swift

Existential types in Swift allow you to store any value that conforms to a specific protocol within a container. However, existential types are less efficient when compared to concrete or generic types. The reason for this inefficiency is that they require dynamic memory allocation, pointer indirection, and dynamic method dispatch (we will look at this more in *Chapter 15, Memory Management*). They can have a noticeable impact on performance and should be used sparingly within performance-critical parts of your application.

To make the impact of existential types clearer, Swift introduced the any keyword to mark existential types. When this was introduced in Swift 5.6, the any keyword was optional, however, starting with Swift 6.0, it is required to use any when working with existential types. Where previously we would write code like this:

```
let mixed: [Dog] = [
    JackRussel(name: "Dash", color: "Brown and White"),
    Mutt(name: "Buddy", color: "Brown")
]
```

Now we write it like this:

```
let mixed: [any Dog] = [
    JackRussel(name: "Dash", color: "Brown and White"),
    Mutt(name: "Buddy", color: "Brown")
]
```

The only difference is the any keyword prior to the Dog type within the array declaration. The any keyword is used to clearly mark existential types, making it obvious when we're using a less efficient existential type instead of a concrete or generic type. This makes it clear that the array contains existential types, which can help us identify potential performance issues within our code.

Now let's look at how Swift can provide synthesized implementations for certain protocols.

Adopting protocols using a synthesized implementation

Swift can automatically provide protocol conformance for the Equatable, Hashable, and Comparable protocols in specific cases. What this means is that we do not need to write all the boilerplate code to implement these protocols; instead, we can use synthesized implementations provided for us. This only works if the structures or enumerations (not classes) contain only stored properties (for structures) or associated values (for enumerations) that also conform to the Equatable, Hashable, and Comparable protocols.

The Equatable, Hashable, and Comparable protocols are provided within the Swift standard library. Any type that conforms to the Equatable protocols can use the equals operator (==) to compare two instances of the type. Any type that uses the Comparable protocol can use comparative operators to compare two instances of the type. Finally, any type that conforms to the Hashable protocol can be hashed to produce an integer hash.

Let's look at one example of this. Let's begin by creating a simple structure that will store names:

```
struct Name {

    var firstName: String
    var lastName: String
}
```

We can now create three instances of the Name structure, as shown in the following code:

```
let name1 = Name(firstName: "Jon", lastName: "Hoffman")
let name2 = Name(firstName: "John", lastName: "Hoffman")
let name3 = Name(firstName: "Jon", lastName: "Hoffman")
```

If we tried to compare the instances of the Name structure, as shown in the following code, we could receive a compile-time error because the Name structure does not conform to the Equatable protocol:

```
name1 == name2
name1 == name3
```

In order to use the comparison operators with instances of the Name structure, the Name structure would need to conform to the Equatable protocol. Normally, we would indicate that it conforms to the Equatable protocol in the structures definition and then provide the required functionality as defined by the protocol. The following code shows how we could make the Name type conform to the Equatable protocol:

```
struct Name: Equatable {
    var firstName = ""
    var lastName = ""
    static func == (lhs: Name, rhs: Name) -> Bool {
        return lhs.firstName == rhs.firstName &&
            lhs.lastName == rhs.lastName

    }
}
```

However, since Swift can automatically provide protocol conformance for the Equatable, Hashable, and Comparable protocols, all that is actually needed is to define that the structure conforms to the Equatable protocol, and the boilerplate code for comparison functionality will be added at compile time.

The following code shows us how this is done:

```
struct Name: Equatable {
    var firstName = ""
    var lastName = ""
}
```

Notice that we do not need the static function, which provided the comparison functionality. We are now able to compare instances of the Name structure as we did in the previous example, but without all the boilerplate code.

Summary

In this chapter, we explored how powerful protocols and protocol extensions are in Swift, beginning with an introduction to protocols as full-fledged types. We saw their importance in providing standard interfaces for multiple types, demonstrated through the use of polymorphism, and examined typecasting using the is and as keywords, illustrating conditional casting and type filtering techniques.

We then looked at protocol extensions, and their role in providing default method and property implementations. We saw how using protocol extensions reduces code duplication and enhances maintainability compared to traditional protocol implementation. The concepts of Any and any were also covered, distinguishing their use in representing instances of any type and marking existential types, and discussing their impact on performance.

Finally, we looked at synthesized implementations for protocols like Equatable, Hashable, and Comparable. This illustrated how Swift can automatically provide conformance for structures and enumerations with stored properties or associated values, eliminating the need for manual implementation of comparison and hashing functionalities.

In the next chapter, we will look at generics and examine how they can be used to write adaptable and reusable code.

4

Generics

For a software developer, the ability to write code that's both adaptable and reusable isn't just a luxury—it's essential. Swift is equipped with numerous features to fulfill this requirement, with generics standing out as one of the most powerful.

While the concept of generics might seem daunting to many developers, they are quite straightforward once you understand them. Generics enable the creation of flexible, reusable functions and types that are compatible with multiple concrete data types. They form the core of many well-known Swift structures, like arrays and dictionaries. By learning about generics, you will not only have a better understanding of the Swift language itself, but you'll be able to write code that is significantly less repetitive and more adaptable to a wide range of programming tasks.

In this chapter, we will cover the following topics:

- What are generics?
- How to create and use generic functions
- How to create and use generic types
- How to use associated types with protocols

Introducing generics

Generics allow us to write very flexible and reusable code that avoids duplication. With a type-safe language, such as Swift, we often need to write functions, classes, and structures that are valid for multiple concrete types. Without generics, we would need to write separate functions for each type we wish to support. With generics, we can write one generic function to provide the functionality for multiple types. Generics allow us to tell a function or type, *"I know Swift is a type-safe language, but I do not know the type that will be needed yet. I will give you a placeholder for now and will let you know what type to enforce later."*

In Swift, we can define both generic functions and generic types. Let's look at generic functions first.

Generic functions

Let's begin by examining the issue that generic functions are designed to solve, and then see how they solve it.

Let's say that we wanted to create functions that swapped the values of two variables; however, for our application, we need functions to swap two `integer` types, two `Double` types, and two `String` types. The following code shows what these functions could look like:

```
func swapInts(a: inout Int,b: inout Int) {
    let tmp = a
    a = b
    b = tmp
}


func swapDoubles(a: inout Double,b: inout Double) {
    let tmp = a
    a = b
    b = tmp
}


func swapStrings(a: inout String, b: inout String) {
    let tmp = a
    a = b
    b = tmp
}
```

With these three functions, we can swap the original values of two `Integer` types, two `Double` types, and two `String` types. Now, let's say, as we develop our application further, we find out that we also need to swap the values of two unsigned `Integer` types, two `Float` types, and even a couple of custom types. We might easily end up with eight or more swap functions. The worst part is that each of these functions contains duplicate code. The only difference between these functions is the parameter types.

We could then use these functions as the following code illustrates.

```
var a = 5
var b = 10
swapInt(a: &a, b: &b)
```

In this code, two variables are set. We then pass those variables to the **swapInt()** function using the & symbol to pass them by reference. This is done so the function can swap the values of the variables. The & symbol is used in conjunction with the **inout** parameter. The **swapDouble()** and the **swapString()** functions could be used in a similar way.

While this solution does work, generics offer a much simpler and more elegant solution that eliminates all the duplicate code.

Defining generic functions

Let's see how we would condense all the swap functions from the previous example into a single generic function:

```
func swapGeneric<T>(a: inout T, b: inout T) {
    let tmp = a
    a = b
    b = tmp
}
```

Let's look at how the swapGeneric function is defined here. The function looks similar to a normal function, except for the capital T. The capital T is a placeholder type that tells Swift that we will be defining the type later. When a type is defined, it will be used in place of the placeholders.

To define a generic function, we include the placeholder type between two angular brackets, <T>, after the function's name. We then use that placeholder type in place of any type definition within the parameter definitions, the return type, or the function itself. The one thing to keep in mind is that, once the placeholder is defined as a type, all the other placeholders assume that type. Therefore, any variable or constant defined with that placeholder must conform to that type.

There is nothing special about the capital T; we could use any valid identifier in place of T. We could use descriptive names, such as key and value, as the Swift language does with dictionaries. The following definitions are perfectly valid:

```
func swapGeneric<G>(a: inout G, b: inout G) {
    //Statements
}
func swapGeneric<xyz>(a: inout xyz, b: inout xyz) {
    //Statements
}
```

In most documentation, generic placeholders are defined with either T (for type) or E (for element). We will, for the purposes of this chapter, continue to use T to define generic placeholders. It is also good practice to use T to define a generic placeholder within our code so that the placeholder is easily recognizable when we are looking at the code later.

> If you do not like using T or E to define generics, you should at least be consistent. I would recommend that you avoid the use of different identifiers to define generics throughout your code.

If we need to use multiple generic types, we can create multiple placeholders by separating them with commas. The following example shows how to define multiple placeholders for a single function:

```
func testGeneric<T,E>(a: T, b: E) {
    //code
}
```

In this example, we are defining two generic placeholders, T and E. In this case, we can set the T placeholder to one type and the E placeholder to a different type.

Calling generic functions

Let's look at how to call a generic function. The following code will swap two integers using the generic swapGeneric() function.

```
var a = 5
var b = 10
swapGeneric(a: &a, b: &b)
print("a:\(a) b:\(b)")
```

If we run this code, the output will be a: 10 b: 5. We can see with this code that we do not have to do anything special to call a generic function. The function infers the type from the first parameter and then sets all the remaining placeholders to that type. Now, if we need to swap the values of two strings, we will call the same function, as follows:

```
var c = "My String 1"
var d = "My String 2"
swapGeneric(a: &c, b: &d)
print("c:\(c) d:\(d)")
```

This function is called in the same way as we called it when we wanted to swap two integers.

One thing that we cannot do is pass two different types into the swap function because we defined only one generic placeholder. If we attempt to run the following code, we will receive an error letting us know that we cannot convert the values of a `String` type to an `Integer` (`Int`) type:

```
var a = 5
var c = "My String 1"
swapGeneric(a: &a, b: &c)
```

The reason the function is looking for an `Int` value is because the first parameter that we pass into the function is an `Int` value, and, therefore, all the generic types in the function become `Int` types.

Now, let's say we have the following function, which has multiple generic types defined:

```
func testGeneric<T,E>(a: T, b: E) {
    print("\(a) \(b)")
}
```

This function would accept parameters of different types; however, since they are of different types, we would be unable to swap the values.

There are additional limitations to generics. For example, we may think that the following generic function is valid:

```
func genericEqual<T>(a: T, b: T) -> Bool {
    return a == b
}
```

However, this code would generate an error because the type of the arguments is unknown at compile time, which means that Swift does not know whether it conforms to the `Comparable` protocol or not. We might think that this is a limit that will make generics hard to use. However, we have a way to tell Swift that we expect the type, represented by a placeholder, to have a certain functionality. This is done with type constraints.

Type constraints

Type constraints specify that a generic type must inherit from a specific class or conform to a particular protocol. This allows us to use the methods or properties defined by the parent class or protocol within the generic function.

Let's look at how to use type constraints by rewriting the `genericEqual` function to use the `Comparable` protocol:

```
func testGenericComparable<T: Comparable>(a: T, b: T) -> Bool {
    return a == b
}
```

To define the type constraint, we put the class or protocol constraint after the generic place-holder, where the generic placeholder and the constraint are separated by a colon. The `testGenericComparable()` function works as we might expect, and it will compare the values of the two parameters and return `true` if they are equal or `false` if they are not.

We can declare multiple constraints just like we declare multiple generic types. The following example shows how to declare two generic types with different constraints:

```
func testFunction<T: MyClass, E: MyProtocol>(a: T, b: E) {
    //Statements
}
```

In this function, the type defined by the `T` placeholder must inherit from the `MyClass` class, and the type defined by the `E` placeholder must conform to the `MyProtocol` protocol.

Now that we have looked at generic functions, let's now turn our attention to generic types.

Generic types

A generic type is a class, structure, or enumeration that can work with any type, just like the way Swift arrays and dictionaries work.

Swift arrays and dictionaries are written so that they can contain any type. The catch is that we cannot mix and match different types within an array or dictionary.

When we create an instance of our generic type, we define the type that the instance will work with. After we define that type, we cannot change the type for that instance.

To demonstrate how to create a generic type, let's create a simple `List` class. This class will use a Swift array as the backend storage and will have the functionality to add items to the list or retrieve values from the list.

Let's begin by seeing how to define our generic List type:

```
class List<T> {

}
```

We can see that we use the <T> tag to define a generic placeholder, as we did when we defined a generic function. This T placeholder can then be used anywhere within the type instead of concrete type definitions.

To create an instance of this type, we would need to define the type of items that our list will hold. The following example shows how to create instances of the generic List type for various types:

```
var stringList = List<String>()
var intList = List<Int>()
var customList = List<MyObject>()
```

The preceding example creates three instances of the List class. The stringList instance can be used with instances of the String type, the intList instance can be used with instances of the integer type, and the customList instance can be used with instances of the MyObject type.

> We are not limited to using generics only with classes. We can also define structures and enumerations as generics. The following example shows how to define a generic structure and a generic enumeration:
>
> ```
> struct GenericStruct<T> {
> }
> ```
>
> ```
>  enum GenericEnum<T> {
> }
> ```

Now let's add the backend storage array to our List class. The items that are stored in this array need to be of the same type that we define when we initiate the class; therefore, we will use the T placeholder for the array's definition. The following code shows the List class with an array named items:

```
class List<T> {
    private var items = [T]()
}
```

This code defines our generic `List` type and uses `T` as the type placeholder so that the array will hold the same types that we defined for the class. We can then use this `T` placeholder anywhere in the class to define the type of an item. That item will then be of the same type that we defined when we created the instance of the `List` class. Therefore, if we create an instance of the `List` type, such as this:

```
var stringList = List<String>()
```

The `items` array will be an array of string instances. If we created an instance of the `List` type like this:

```
var intList = List<Int>()
```

The `items` array will be an array of integer instances.

Now let's create the add() method, which will be used to add an item to the list. We will use the `T` placeholder within the method declaration to define that the parameter will be of the same type that we declared when we initiated the class:

```
func add(item: T) {
    items.append(item)
}
```

To create a standalone generic function, we add the `<T>` declaration after the function name to declare that it is a generic function; however, when we use a generic method within a generic type, we do not need the `<T>` declaration. Instead, we just need to use the type that we defined in the class declaration. If we wanted to introduce another generic type, we could define it with the method declaration.

Now, let's add the getItemAtIndex() method, which will return the item from the backend array at the specified index:

```
func getItemAtIndex(index: Int) -> T? {
    if items.count>index {
        return items[index]
    } else {
        return nil
    }
}
```

The getItemAtIndex() method accepts one argument, which is the index of the item we want to retrieve. We then use the T placeholder to specify that our return type is an optional that might be of the T type or that might be nil. If the backend storage array contains an item at the specified index, we will return that item; otherwise, we return nil.

Here is the entire generic List class:

```
class List<T> {
    private var items = [T]()

    func add(item: T) {
        items.append(item)
    }

    func getItemAtIndex(index: Int) -> T? {
        guard items.count < index else {
            return items[index]
        }
        return nil
    }
}
```

Now, let's look at how to use the List class. The following code shows how to use the List class to store instances of the String type:

```
var list = List<String>()
list.add(item: "Hello")
list.add(item: "World")

print(list.getItemAtIndex(index: 1)!)
```

In the preceding code, we start off by creating an instance of the List type and indicate that it would store instances of the String type. We then use the add() method twice to store two items in the list. Finally, we use the getItemAtIndex() method to retrieve the item at index number 1, which will display World to the console.

We can also define our generic types with multiple placeholder types, similar to how we use multiple placeholders in our generic functions. To use multiple placeholder types, we separate them with commas. The following example shows how to define multiple placeholder types:

```
class MyClass<T,E>{
//Code
}
```

We then create an instance of the MyClass type that uses instances of the String and Int types, like this:

```
var mc = MyClass<String, Int>()
```

We can also use type constraints with generic types. Once again, using a type constraint for a generic type is exactly the same as using one with a generic function. The following code shows how to use a type constraint to ensure that the generic type conforms to the Comparable protocol:

```
class MyClass<T: Comparable>{}
```

So far in this chapter, we have seen how to use placeholder types with functions and types. Now, let's see how we can conditionally add extensions to a generic type.

Conditionally adding extensions with generics

We can add extensions to a generic type conditionally if the type conforms to a protocol. For example, if we wanted to add a sum() method to our generic List type only if the generic type conforms to the Numeric protocol, we could do the following:

```
extension List where T: Numeric {
    func sum () -> T {
        return items.reduce (0, +)


    }
}
```

This extension adds the sum() method to any list instance where the T type conforms to the Numeric protocol. This means that the list instance in the previous example, where the list was created to hold String instances, would not receive this method.

In the following code, where we create an instance of the List type that contains integers, the instance will receive sum() and can be used as shown:

```
var list2 = List<Int>()
list2.add(item: 2)
list2.add(item: 4)
list2.add(item: 6)
print(list2.sum())
```

The output of this is 12.

We are also able to conditionally add functions inside a generic type or extension.

Conditionally adding functions

Conditionally adding extensions, as we saw in the last section, works well; however, if we had separate functionality that we wished to add for different conditions, we would have to create separate extensions for each condition.

In Swift, we are able to conditionally add functions to a generic type or extension. Let's take a look at this by rewriting the extension in the previous section:

```
extension List {
    func sum () -> T where T: Numeric {
        items.reduce (0, +)
    }
}
```

With this code, we moved the where T: Numeric clause out of the extensions declaration and into the function declaration. This will conditionally add the sum function if the type conforms to the Numeric protocol. Now we can add additional functions to the extension with different conditions, as shown in the following code:

```
extension List {
    func sum () -> T where T: Numeric {
        items.reduce (0, +)
    }
    func sorted() -> [T] where T: Comparable {
        items.sorted()
    }
}
```

In the preceding code, we added an additional function named sorted() that will only be applied to instances where the type conforms to the Comparable protocol.

This enables us to put functions with different conditions in the same extension or generic type rather than creating multiple extensions. I would recommend conditionally adding functions, as shown in this section, over conditionally adding extensions, as shown in the previous section.

Now, let's look at conditional conformance.

Conditional conformance

Conditional conformance allows a generic type to conform to a protocol only if the type meets certain conditions. For example, if we wanted our List type to conform to the Equatable protocol only if the type stored in the list also conformed to the Equatable protocol, we could use the following code:

```
extension List: Equatable where T: Equatable {
    static func == (l1:List, l2:List) -> Bool {
        if l1.items.count != l2.items.count {
            return false
        }
        for (e1, e2) in zip(l1.items, l2.items) {

            if e1 != e2 {
                return false
            }
        }
        return true
    }
}
```

This code will add conformance to the Equatable protocol to any instance of the List type where the type that is stored in the list also conforms to the Equatable protocol.

There is also a new function shown here that we have not talked about: the zip() function. This function will loop through two sequences, in our case arrays, simultaneously, and create pairs (e1 and e2) that we can compare.

The comparison function will first check to see that each array contains the same number of elements and, if not, it will return `false`. It then loops through each array, simultaneously comparing the elements of the arrays; if any of the pairs do not match, it will return `false`. If all previous tests pass, `true` is returned, which indicates that the `list` instances are equal because the elements in the list are the same.

Now, let's see how we can add generic subscripts to a non-generic type.

Generic subscripts

In Swift, we can create generic subscripts, where either the subscript's return type or its parameters may be generic. This gives our subscripts the same flexibility and type safety that we get with generic types and functions.

Let's look at how we can create a generic subscript. In this first example, we will create a subscript that will accept one generic parameter:

```
Struct HashProvider {
    subscript<T: Hashable>(item: T) -> Int {
        return item.hashValue
    }
}
```

When we create a generic subscript, we define the placeholder type after the `subscript` keyword. In the previous example, we define the `T` placeholder type and use a type constraint to ensure that the type conforms to the `Hashable` protocol. This will allow us to pass in an instance of any type that conforms to the `Hashable` protocol.

As we mentioned at the start of this section, we can also use generics for the return type of a subscript. We define the generic placeholder for the return type exactly as we did for the generic parameter. The following example illustrates this:

```
subscript<T>(key: String) -> T? {
    return dictionary[key] as? T
}
```

In this example, we define the `T` placeholder type after the `subscript` keyword, as we did in the previous example. We then use this type as our return type for the `subscript`.

Now, let's look at what associated types are and how to use them.

Associated types

An associated type declares a placeholder name that can be used instead of a concrete type within a protocol where the type to use is not specified until the protocol is adopted. When creating generic functions and types, we used a very similar syntax, as we have seen throughout this chapter. Defining associated types for a protocol, however, is very different. We specify an associated type using the associatedtype keyword.

Let's see how to use associated types when we define a protocol. In this example, we will define the QueueProtocol protocol, which defines the capabilities that need to be implemented by the queue that implements it:

```
protocol QueueProtocol {
    associatedtype QueueType
    mutating func add(item: QueueType)
    mutating func getItem() -> QueueType?
    func count() -> Int
}
```

In this protocol, we defined one associated type, named QueueType. We then used this associated type twice within the protocol: once as the parameter type for the add() method and once as an optional, when we defined the return type of the getItem() method.

Any type that implements the Queue protocol must be able to specify the type to use for the QueueType placeholder and must also ensure that only items of that type are used where the protocol uses the QueueType placeholder.

Let's look at how to implement Queue in a non-generic class called IntQueue. This class will implement the Queue protocol using the integer type:

```
class IntQueue: QueueProtocol {
    var items = [Int]()
    func add(item: Int) {
        items.append(item)
    }
    func getItem() -> Int? {
        return items.count > 0 ? items.remove(at: 0) : nil
    }
    func count() -> Int {
```

```
            return items.count
    }
}
```

In the IntQueue class, we begin by defining our backend storage mechanism as an array of integer types. We then implement each of the methods defined in the QueueProtocol protocol, replacing the QueueType placeholder defined in the protocol with the Int type. In the add() method, the parameter type is defined as an instance of Int type, and in the getItem() method, the return type is defined as an optional that might return an instance of Int type or nil.

We use the IntQueue class as we would use any other class. The following code shows this:

```
var intQ = IntQueue()
intQ.add(item: 2)
intQ.add(item: 4)

print(intQ.getItem()!)
intQ.add(item: 6)
```

This code begins by creating an instance of the IntQueue class, named intQ. We then call the add() method twice to add two values of the integer type to the intQ instance. We then retrieve the first item in the intQ instance by calling the getItem() method. This line will print the number 2 to the console. The final line of code adds another instance of the integer type to the intQ instance.

In the preceding example, we implemented the Queue protocol in a non-generic way. This means that we replaced the placeholder types with an actual type. We can also implement Queue with a generic type. Let's see how we would do this:

```
class GenericQueue<T>: Queue {
    var items = [T]()
    func add(item: T) {
        items.append(item)
    }
    func getItem() -> T? {
        return items.count > 0 ? items.remove(at:0) : nil
    }
    func count() -> Int {
        return items.count
    }
}
```

As we can see, the GenericQueue implementation is very similar to the IntQueue implementation, except that we define the type to use as the generic placeholder T. We can then use the
GenericQueue class as we would use any generic class. Let's see how we would use it:

```
var intQ2 = GenericQueue<Int>()
intQ2.add(item: 2)
intQ2.add(item: 4)
print(intQ2.getItem()!)
intQ2.add(item: 6)
```

We begin by creating an instance of the GenericQueue class that will use the Int type and name
it intQ2. Next, we call the add() method twice to add two instances of the integer type to the
intQ2 instance. We then retrieve the first item in the queue that was added using the getItem()
method and print the value to the console. This line will print the number 2 to the console.

We can also use type constraints with associated types. When the protocol is adopted, the type
defined for the associated type must inherit from the class or conform to the protocol defined by
the type constraint. The following line defines an associated type with a type constraint:

```
associatedtype QueueType: Hashable
```

In this example, we specify that when the protocol is implemented, the type defined for the associated type must conform to the Hashable protocol. Now let's look at implicitly opened existentials,
which enable us to use protocols that contain associated types as a type.

Implicitly opened existentials

Prior to Swift 5.7, when SE-0352 was added, it was not possible to use protocols that contained
associated types or self requirements as types. This made working with protocols that contained
associated types cumbersome at times. Developers often had to resort to using type erasure techniques or additional boilerplate code to make their generic protocols usable as types. This was
because Swift's type system didn't allow a protocol with associated types to be used as a type
without specifying what those associated types were. Let's look at an example of this. To begin
with, let's create a Drawable protocol with two types that conform to it:

```
protocol Drawable {
    func draw()
}

struct Circle: Drawable {
```

```
        func draw() {
            print("Drawing a circle")
        }
    }

    struct Square: Drawable {
        func draw() {
            print("Drawing a square")
        }
    }
```

In this code, we define the Drawable protocol and then create two structures, Circle and Square that conform to it. Prior to SE-0352, the following function would not be valid:

```
    func drawAll(_ items: [Drawable]) {
        for item in items {
            item.draw()
        }
    }
```

Now, with the addition of the existential any and SE-0352, we can write this code like this:

```
    func drawAll(_ items: [any Drawable]) {
        for item in items {
            item.draw()
        }
    }
```

Notice that we added the any keyword in the parameter definition. SE-0352 allows the Swift compiler to implicitly open the existential type. This means that the compiler can now infer the underlying concrete type of the existential and pass it to the generic function without the need for explicit type annotations or conversions.

Summary

Generic types can be incredibly useful, and they are also the basis of the Swift standard collection types (arrays and dictionaries); however, we have to be careful to use them correctly.

We saw a couple of examples in this chapter that show how generics can make our lives easier. The `swapGeneric()` function that was shown at the beginning of the chapter is a good use of a generic function because it allows us to swap two values of any type we choose, while only implementing the swap code once.

The generic `List` type is also a good example of how to make custom collection types that can be used to hold any type. The way that we implemented the generic `List` type in this chapter is similar to how Swift implements an array and dictionary with generics. We also covered generic subscripts and associated types.

In the next chapter, we will look at the differences between value and reference types.

5

Value and Reference Types

Understanding the distinction between value and reference types is essential to writing efficient and easy-to-manage code in Swift. In Swift, every piece of data you work with falls into one of these two categories, each with its own unique behavior and implications for memory management, mutability, and performance.

In this chapter, we will explore the concepts of value and reference types in depth, uncovering their characteristics, differences, and best practices for utilizing them effectively. By understanding the differences and similarities between value and reference types, we can make informed decisions when designing our Swift applications, ensuring we are optimizing our code for performance and memory management while having a clean and easy-to-maintain code base.

In this chapter, you will explore the following topics:

- The differences between value types and reference types
- Why recursive data types cannot be created as a value type
- How dynamic dispatch works
- How to implement copy-on-write in your custom type

In Swift, we have the flexibility to define our own custom types, which can be categorized as either reference types or value types. It's essential to understand the differences between these categories as they significantly impact our choice of type for custom implementations. Let's start examining the differences between these two types.

Value types and reference types

Structures are an example of value types. When structures are passed as a parameter to a function or assigned to a new variable, a copy of the structure is used rather than the original value. This means that each part of our code receives its own copy of the structure, enabling it to make changes as necessary without impacting the original instance of the structure.

> We are using structures for our examples here, but enumerations are also value types and function in the same way.

Classes are reference types; therefore, when we pass an instance of a class as a parameter to a function or assign it to a new variable, we are using a reference to the original instance of the class. Any changes made to the instance of the class will persist.

It is very important to understand the difference between value types and reference types. We will discuss a very high-level view here, but will provide additional details in *Chapter 15, Memory Management*.

To illustrate the difference between value types and reference types, let's examine a real-world object: a book. If we have a friend who wants to read *Mastering Swift 6*, we could either buy them their own copy or share ours.

If we bought our friend their own copy of the book, then any notes they make in the book would remain in their copy of the book and would not be reflected in our copy. This is how passing by value works with value types such as structures and enumerations. Any changes that are made to the value type within the function are not reflected back to the original instance.

If we share our copy of the book, then any notes that our friend makes in the book will remain in the book when it is returned to us. This is how passing by reference works. Any changes that are made to the instance of the class remain when the function exits.

> When we say passing an instance of a value type, it implies that a copy of the instance needs to be made. With this realization, you might have concerns regarding the performance implications, especially with sizable value types moving between different segments of our code.
>
> To address this, for structures with the potential of growing significantly in size, we can implement copy-on-write mechanisms. We will see how to implement copy-on-write later in this chapter.

In this section, we are going to examine the differences between value types and reference types, so we understand when to use each type. Let's begin by creating two types: one is going to be a structure (or value type) and the other is going to be a class (or reference type).

The first type that we will examine is named GradeValueType. We will implement this type using a structure, which means it is a value type, as its name suggests:

```
struct GradeValueType {
    var name: String
    var assignment: String
    var grade: Int
}
```

♡ **Quick tip**: Enhance your coding experience with the **AI Code Explainer** and **Quick Copy** features. Open this book in the next-gen Packt Reader. Click the **Copy** button (1) to quickly copy code into your coding environment, or click the **Explain** button (2) to get the AI assistant to explain a block of code to you.

```
function calculate(a, b) {
    return {sum: a + b};
};
```

Copy   Explain
 1   2

⊟ **The next-gen Packt Reader** is included for free with the purchase of this book. Scan the QR code OR go to packtpub.com/unlock, then use the search bar to find this book by name. Double-check the edition shown to make sure you get the right one.

In GradeValueType, we define three properties. Two of the properties are of the String type (name and assignment), and one is of the Integer type (grade). Now, let's take a look at how we can implement this as a class:

```
class GradeReferenceType {
    var name: String
    var assignment: String
    var grade: Int
    init(name: String, assignment: String, grade: Int) {
        self.name = name
        self.assignment = assignment
        self.grade = grade
    }
}
```

GradeReferenceType defines the same three properties as in GradeValueType; however, we need to define an initializer in GradeReferenceType that we did not need to define in GradeValueType. The reason for this is that structures provide us with a default initializer that will initialize all the properties that need to be initialized if we do not provide an initializer.

Let's take a look at how we can use each of these types. The following code shows how we can create instances of each of these types:

```
var ref = GradeReferenceType(name: "Jon", assignment: "Math Test 1",
                            grade: 90)
var val = GradeValueType(name: "Jon", assignment: "Math Test 1",
                         grade: 90)
```

As you can see in this code, instances of structures are created in the same way as instances of classes. Being able to use the same format to create instances of structures and classes is good because it makes our lives easier; however, we do need to bear in mind that value types behave in a different manner to reference types.

Let's explore this; the first thing we need to do is create two functions that will change the grades for the instances of the two types:

```
func extraCreditReferenceType(ref: GradeReferenceType, extraCredit: Int) {
    ref.grade += extraCredit
}
func extraCreditValueType(val: GradeValueType, extraCredit: Int) {
```

```
        val.grade += extraCredit
    }
```

Each of these functions takes an instance of one of our types and an extra credit amount. Within the function, we add the extra credit amount to the grade.

If we try to use this code, we receive an error in the extraCreditValueType() function telling us that the left side of the mutable operation is not mutable. The reason for this is that a value type parameter, by default, is immutable because the function is receiving an immutable copy of the parameter and not a reference to the original.

Using a value type like this protects us from making accidental changes to the instances; this is because the instances are scoped to the function or type in which they are created. Value types also protect us from having multiple references to the same instance. Therefore, they are, by default, thread (concurrency) safe because each thread will have its own version of the value type. If we absolutely need to change an instance of a value type outside of its scope, we could use an inout parameter.

inout parameters and value types

We define inout parameters by placing the inout keyword at the start of the parameter's definition. An inout parameter has a value that is passed into the function. This value is then modified by the function and is passed back out when the function exits to replace the original value.

Let's explore how we can use an inout parameter. We will begin by creating a function that is designed to retrieve the grade for an assignment from a data store. However, to simplify our example, we will simply generate a random score.

The following code demonstrates how we can write this function:

```
func getGradeForAssignment(assignment: inout GradeValueType) {
    // Code to get grade from DB
    // Random code here to illustrate
    let num = Int.random(in: 80..<100)
    assignment.grade = num
    print("Grade for \(assignment.name) is \(num)")
}
```

This function is designed to retrieve the grade for the assignment that is defined in the GradeValueType instance and is then passed into the function. Once the grade is retrieved, we will use it to set the grade property of the GradeValueType instance. We will also print the grade out to the console so that we can see what grade it is.

Now, let's look at how to use this function:

```swift
var mathGrades = [GradeValueType]()
let students = ["Jon", "Kailey", "Kai"]
var mathAssignment = GradeValueType(name: "", assignment: "Math
                                    Assignment", grade: 0)

for student in students {
    mathAssignment.name = student
    getGradeForAssignment(assignment: &mathAssignment)
    mathGrades.append(mathAssignment)
}

for assignment in mathGrades {
    print("\(assignment.name): grade \(assignment.grade)")
}
```

In the previous code, we created a mathGrades array that will store the grades for our assignment, and a students array that will contain the names of the students for whom we wish to retrieve the grades. We then created an instance of the GradeValueType structure that contains the name of the assignment. We will use this instance to request the grades from the getGradeForAssignment() function.

Notice that when we pass in the mathAssignment instance, we prefix the name of the instance with the & symbol. This lets us know that we are passing the reference to the original instance and not a copy.

Now that everything is defined, we will loop through the list of students to retrieve the grades. The output of this code will look similar to the following snippet:

```
Grade for Jon is 87
Grade for Kailey is 90
Grade for Kai is 83
Jon: grade 87
Kailey: grade 90
Kai: grade 83
```

The output from this code is what we expected to see, where each instance in the mathGrades array represents the correct grade. The reason this code works correctly is that we are passing a reference from the mathAssignment instance to the getGradeForAssignment() function, and not a copy.

Sometimes, when working with value types, we prefer not to create new copies of instances when passing them. This is where noncopyable types come in.

Noncopyable types

In Swift, all structs, classes, enums, generic type parameters, and protocols automatically conform to the Copyable protocol. This protocol does not require the implementation of any methods or properties, and typically, there's no need to explicitly declare conformance. The Copyable protocol allows for the creation of multiple identical copies of instances of these types.

To prevent a type from being copyable, we can opt out by using ~Copyable. Types that opt out of this protocol are referred to as noncopyable types. These noncopyable types enable us to create values with unique ownership and prevent instances from being copied. This approach allows single instances of value types, such as structs and enums, to be shared across different parts of the code while maintaining a single owner.

Let's look at how we could create a noncopyable type:

```
struct Person: ~Copyable {
    var firstName: String
    var lastName: String
    var emailAddress: String
}
```

In this code, using ~Copyable makes the Person type noncopyable. Noncopyable types introduce the idea of unique ownership. When a noncopyable type is passed within our code, we either allow another part of the code to temporarily borrow the instance or we transfer ownership of (consume) the instance, instead of creating a new copy. Let's look at examples of these concepts:

```
func sendEmail(_ user: borrowing Person) {
    print("Sending Email to \(user.firstName) \(user.lastName)")
}

func consumeUser(_ user: consuming Person) {
    print("Consuming User")
}
```

The sendEmail() function uses the borrowing keyword to borrow an instance of the Person type, while the consumeUser() function uses the consuming keyword to transfer ownership of an instance of the Person type.

Borrowing an instance grants temporary access to a value without consuming it. There are three key concepts to remember when borrowing instances:

- Borrowing allows read-only access to a noncopyable value and does not transfer ownership.
- Borrowing is useful when you need to inspect or use a value without consuming it.
- Borrowed values are thread-safe and can be accessed concurrently by multiple parts of your code.

When we **consume** an instance, we are transferring ownership from one part of our code to another. Three key concepts to remember when consuming instances are:

- Consuming transfers ownership of a noncopyable value, and the original variable becomes invalid.
- Global instances of noncopyable types cannot be consumed.
- Consuming methods, such as the consumeUser() method, end the lifetime of the object when the method returns.

Let's explore how these two functions work and how ownership is managed. Since we cannot consume a global instance of a noncopyable type, we'll create a new function to demonstrate, as shown here:

```
func userFunction() {
    let user = Person(firstName: "Jon", lastName: "Hoffman",
                      emailAddress: "Jon@mydomain.com")

    sendEmail(user)
    consumeUser(user)
}
```

In this function, we create a new instance of the Person type. We then call the sendEmail() function, which borrows the instance to send an email. After that, we call the consumeUser() function, which takes ownership of the user instance, rendering the original user constant invalid.

If we were to reverse the order of these two function calls, as shown in the following code, we would encounter an error indicating that the instance was used after it was consumed, and the code would fail to run:

```
func userFunction() {
    let user = Person(firstName: "Jon", lastName: "Hoffman",
                      emailAddress: "Jon@mydomain.com")
    consumeUser(user)
    sendEmail(user
}
```

This code is invalid because the consumeUser() function takes ownership of the instance, making the user instance invalid after the function is called.

With noncopyable types, we can create consuming methods that invalidate the instance once the method has been executed. For example, if we're creating a secret message that should disappear after being read, we could achieve this with the following code:

```
struct SecretMessage: ~Copyable {
    private var message: String
    init(_ message: String) {
        self.message = message
    }

    consuming func read() {
        print("\(message)")
    }
}
```

In this code, notice that the read() method is marked with the consuming keyword; therefore, once this method is executed, the instance will no longer be valid. Let's see how this works with the following code:

```
func secretMessageFunction() {
    var secretMessage = SecretMessage("My Message")

    secretMessage.read()
}
```

In this code, we create an instance of the SecretMessage type and then call the read() method. After the read() method is called, if we tried to access the instance again, such as calling the read() method again, we would encounter an error indicating that the instance was used after it was consumed.

An additional thing to note is that noncopyable types can conform to protocols but only if those protocols are also marked as noncopyable. For example, our SecretMessage type could conform to the following protocol because it is also marked as noncopyable:

```
protocol Message: ~Copyable {
    consuming func read()
}
```

Noncopyable types can enhance performance in several key areas:

- By eliminating unnecessary copy of data, memory management becomes more efficient.
- The compiler can more accurately track an instance's lifecycle.
- Resource leaks are reduced when, as examples, file descriptions, and network sockets are wrapped in non-copyable types, ensuring they cannot be accidentally duplicated.
- By enforcing unique ownership, noncopyable types can reduce locks in concurrent code.

Noncopyable types are powerful tools that help us enforce ownership and manage resources.

There are some things that can't be done with value types that we can do with reference types. The first thing that we will look at is the recursive data type.

Recursive data types for reference types

A recursive data type is a type that contains values of the same type as a property of the type. Recursive data types are used when we want to define dynamic data structures, such as lists and trees. The size of these dynamic data structures can grow or shrink, depending on our runtime requirements.

Linked lists are great examples of dynamic data structures made possible using a recursive data type. Essentially, a linked list is a collection of nodes that are connected together, where each node has some data and a link to the next node in line. If a node loses its link to the next one, the rest of the list becomes inaccessible because each node only knows about its immediate neighbor.

Some linked lists also allow for backward movement by keeping track of both previous and next nodes, enabling navigation in both directions. *Figure 5.1* shows how a very basic linked list works:

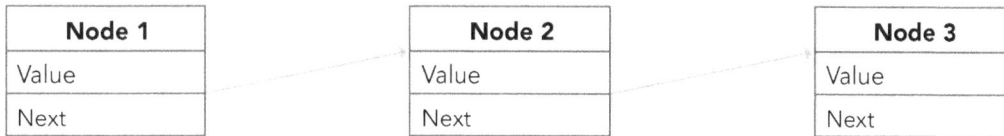

Figure 5.1: A basic linked list

The following code shows how we can create a linked list using a reference type:

```
class LinkedListReferenceType {
    var value: String
    var next: LinkedListReferenceType?
    init(value: String) {
        self.value = value
    }
}
```

In the LinkedListReferenceType class, we have two properties. The first property is named value, and it contains the data for this instance. The second property is named next, which points to the next item in the linked list. If the next property is nil, then this instance will be the last node in the list.

If we try to implement this linked list as a value type, the code will be similar to the following:

```
struct LinkedListValueType {
    var value: String
    var next: LinkedListValueType?
}
```

When we add this code to a playground, we receive an error telling us that Swift does not allow recursive value types. However, we can implement them as a reference type, as we showed earlier.

If you think about it, recursive value types are a really bad idea because of how value types work. Let's examine this for a minute, because it will really emphasize the difference between value types and reference types. It will also help you to understand *why* we need reference types.

Let's say that we are able to create the `LinkedListValueType` structure without any errors. Now, let's create three nodes for our list, as shown in the following code:

```
var one = LinkedListValueType(value: "One",next: nil)
var two = LinkedListValueType(value: "Two",next: nil)
var three = LinkedListValueType(value: "Three",next: nil)
```

Now, we will link these nodes together using the following code:

```
one.next = two
two.next = three
```

Do you see the problem with this code? If not, think about how a value type is passed. In the first line, one.next = two, we are not actually setting the next property to the original two instance; in fact, we are actually setting it to a copy of the two instance, because by implementing `LinkedListValueType` as a value type, we are passing the value and not the actual instance. This means that in the next line, two.next = three, we are setting the next property of the original two instance to the three instance.

However, this change is not reflected back in the copy that was made for the next property of the one instance. Sounds a little confusing? Let's clear it up a little by looking at a diagram that shows the state of our three `LinkedListValueType` instances if we were able to run this code:

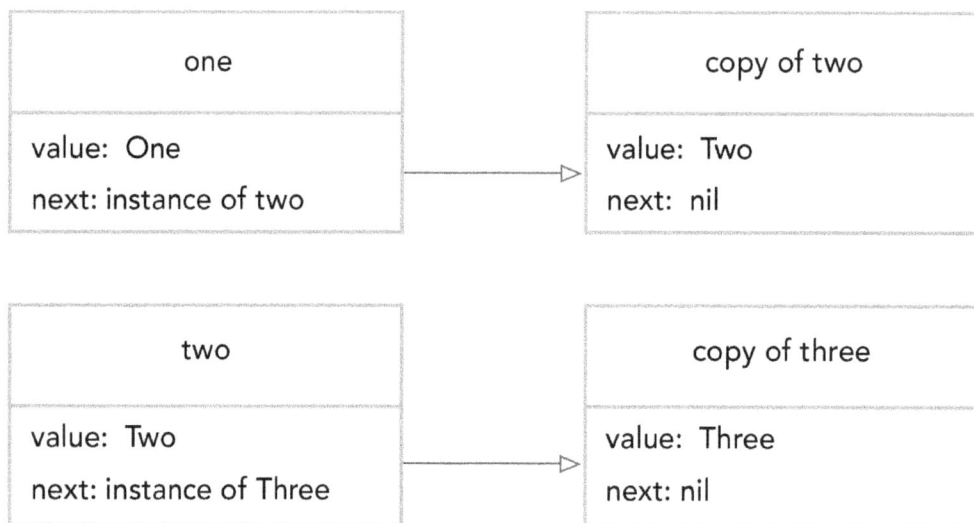

Figure 5.2: A linked list structure example

As you can see from the diagram, the next property of the one instance is pointing to a copy of the two instance, whose next property is still nil. The next property of the original two instance, however, is pointing to the three instance. This means that if we try to go through the list by starting at the one instance, we will not reach the three instance because the copy of the two instance will still have a next property that is nil.

Another thing that we can only do with reference (or class) types is class inheritance.

Inheritance for reference types

In object-oriented programming, inheritance refers to one class (known as a **sub** or **child class**) being derived from another class (known as a **super** or **parent class**). These subclasses will inherit methods, properties, and other characteristics from their superclass. With inheritance, we can also create a class hierarchy where we can have multiple layers of inheritance.

Let's take a look at how we can create a class hierarchy with classes in Swift. We will start by creating a base class named Animal:

```
class Animal {
    var numberOfLegs = 0
    func sleeps() {
        print("zzzzz")
    }
    func walking() {
        print("Walking on \(numberOfLegs) legs")
    }
    func speaking() {
        print("No sound")
    }
}
```

In the Animal class, we define one property (numberOfLegs) and three methods (sleeps(), walking(), and speaking()). By defining these in the Animal class, any class that is a subclass of the Animal class will also have these properties and methods.

Let's examine how this works by creating two classes that are subclasses of the Animal class. These two classes will be named Biped (an animal with two legs) and Quadruped (an animal with four legs):

```
class Biped: Animal {
    override init() {
        super.init()
        numberOfLegs = 2
    }
}
class Quadruped: Animal {
    override init() {
        super.init()
        numberOfLegs = 4
    }
}
```

Since these two classes inherit all the properties and methods from the Animal class, all we need to do is create an initializer that sets the numberOfLegs property to the correct number of legs.

Now, let's add another layer of inheritance by creating a Dog class that will be a subclass of the Quadruped class:

```
class Dog: Quadruped {
    override func speaking() {
        print("Barking")
    }
}
```

In the Dog class, we inherit from the Quadruped class, which, in turn, inherits from the Animal class. Therefore, the Dog class will have all the properties, methods, and characteristics of both the Animal and Quadruped classes. If the Quadruped class overrides anything from the Animal class, then the Dog class will inherit the version from the Quadruped class.

We can create very complex class hierarchies in this manner; for example, *Figure 5.3* expands on the class hierarchy that we just created to add several other Animal classes:

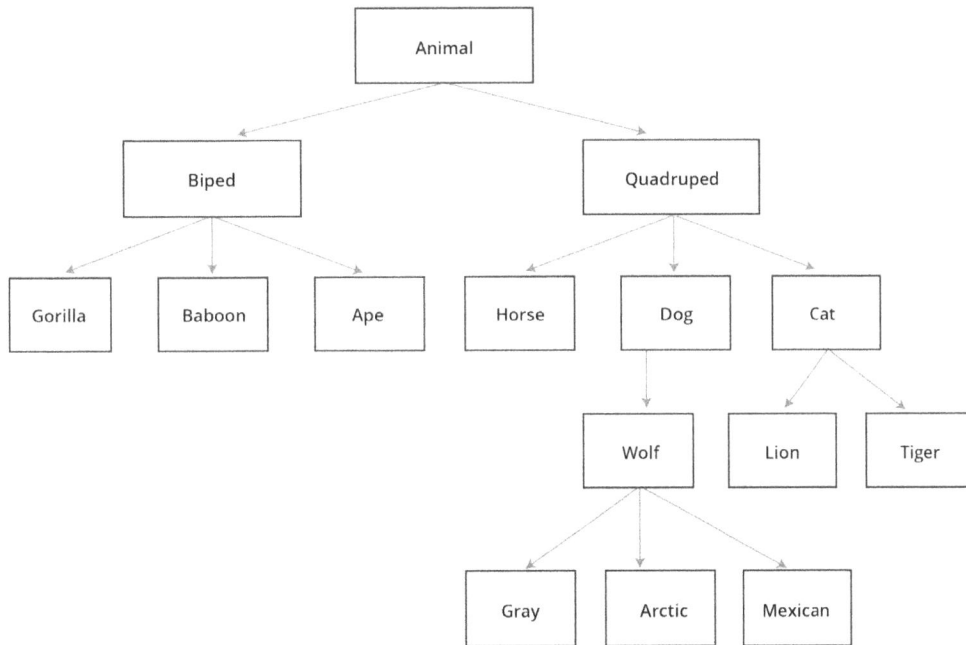

Figure 5.3: Animal class hierarchy

Class hierarchies can get very complex. However, as you just saw, they can eliminate a lot of duplicate code because our subclasses inherit methods, properties, and other characteristics from their superclasses. Because of this, we do not need to recreate them in each of the subclasses.

The biggest drawback of a class hierarchy is the complexity. When we have a complex hierarchy (as shown in the preceding diagram), it is easy to make a change and not realize how it is going to affect each of the subclasses. If you consider the Dog and Cat classes, for example, we may want to add a furColor property to our Quadruped class so that we can set the color of the animal's fur. However, horses do not have fur; they have hair. Therefore, before we can make any changes to a class in our hierarchy, we need to understand how it will affect all the subclasses in the hierarchy.

In Swift, it is best to avoid using complex class hierarchies (as shown in this example) and instead use a protocol-oriented design, unless, of course, there are specific requirements to use them. We will look at how to use protocol-oriented design paradigms in *Chapter 20, Protocol-Oriented Design*.

Now that we have a good understanding of reference and value types, let's explore dynamic dispatch and how it works with our class hierarchy.

Dynamic dispatch

In the previous section, we learned how to use inheritance with classes to inherit and override the functionality defined in a superclass. You may be wondering how and when the appropriate implementation is chosen. The process of choosing which implementation to use is performed at runtime and is known as **dynamic dispatch**.

One of the key points from the last paragraph is that the implementation is chosen at runtime. What this means is that a certain amount of runtime overhead is associated with using class inheritance, as shown in the previous *Inheritance for reference types* section. For most applications, this overhead is not a concern; however, for performance-sensitive applications such as games, this overhead can be costly.

In Swift, dynamic dispatch is implemented using a virtual method table, or VTable. Each class has its own VTable, which stores pointers to the implementations of its overridden methods. When a virtual method is called on an object, the runtime uses the object's class to look up the corresponding method pointer in the VTable, and then calls that implementation. Here's how the process works in more detail:

1. **Object initialization**: When an object is created, the runtime will allocate memory for the object and initialize its instance variables. It also sets up the object's VTable, which contains pointers to the implementations of the object's overridden methods.

2. **Method dispatch**: When a method is called on an object, the runtime will begin by looking up the object's class to find the corresponding VTable. It will then use the method name to retrieve the pointer to the correct method implementation.

3. **Indirect call**: Finally, the runtime will perform an indirect call to the method implementation pointed to by the VTable entry. This indirect call is slightly slower than a direct call, as it involves an extra level of indirection.

One of the ways that we can reduce the overhead associated with dynamic dispatch is to use the final keyword. The final keyword puts a restriction on the class, method, or function, which indicates that it cannot be overridden, in the case of a method or function, or subclassed, in the case of a class.

To use the `final` keyword, you put it before the class, method, or function declaration, as shown in the following code:

```
final func myFunc() {}
final var myProperty = 0
final class MyClass {}
```

In the *Inheritance for reference types* section, we defined a `class` hierarchy that started with the `Animal` superclass. If we want to restrict subclasses from overriding the `walking()` method and the `numberOfLegs` property, we can change the `Animal` implementation, as shown in the next example:

```
class Animal {
    final var numberOfLegs = 0
    func sleeps() {
        print("zzzzz")
    }
    final func walking() {
        print("Walking on \(numberOfLegs) legs")
    }
    func speaking() {
        print("No sound")
    }
}
```

This change allows the application, at runtime, to make a direct call to the `walking()` method rather than an indirect call, which gives the application a slight performance boost. If you must use a class hierarchy, it is good practice to use the `final` keyword wherever possible; however, it is better to use a protocol-oriented design, with value types, to avoid this if possible.

Now, let's take a look at something that can help with the performance of our custom value types, copy-on-write.

Copy-on-write

Normally, when we pass an instance of a value type, such as a structure, a new copy of the instance is created. This means that if we have a large data structure that contains, as an example, 1,000,000 elements, then every time we pass that instance, we will have to copy all 1,000,000 elements. This can have a detrimental impact on the performance of our applications, especially if we pass the instance to numerous functions.

To solve this issue, Apple has implemented the copy-on-write feature for all the data structures, such as Array, Dictionary, and Set, in the Swift standard library. With copy-on-write, Swift does not make a second copy of the data structure until a change is made to that data structure. Therefore, if we pass an array of 1,000,000 elements to another part of our code, and that code does not make any changes to the array, we will avoid the runtime overhead of copying all elements.

This is a very useful feature and can greatly increase the performance of our applications. However, our custom value types do not automatically get this feature by default. In this section, we will explore how we can use reference types and value types together to implement the copy-on-write feature for our custom value types. To do this, we will create a very basic queue type that will demonstrate how you can add copy-on-write functionality to your custom value types.

Creating a backend storage type

We will start by creating a backend storage type called BackendQueue and will implement it as a reference type. The following code gives our BackendQueue type the basic functionality of a queue type:

```
fileprivate class BackendQueue<T> {
    private var items = [T]()
    public func addItem(item: T) {
        items.append(item)
    }
    public func getItem() -> T? {
        if items.count > 0 {
            return items.remove(at: 0)
        } else {
            return nil
        }
    }
    public func count() -> Int {
        return items.count
    }
}
```

The BackendQueue type is a generic type that uses an array as the datastore. This type contains three methods, which enable us to add items to the queue, retrieve an item from the queue, and return the number of items in the queue. We use the fileprivate access level to prevent the use of this type outside of the source file it is defined in, because it should only be used to implement the copy-on-write feature for our main queue type.

We now need to add a couple of extra items to the BackendQueue type to implement the copy-on-write feature for the main queue type. The first thing that we will add is a public default initializer and a private initializer that can be used to create a new instance of the BackendQueue type; the following code shows the two initializers:

```
public init() {}

private init(_ items: [T]) {
    self.items = items
}
```

The public initializer will be used to create an instance of the BackendQueue type without any items in the queue. The private initializer will be used to create a new instance of the BackendQueue type with the items array populated. This will be used internally to create a copy of an existing BackendQueue type.

Now, we will need to create a method that will use the private initializer to create a copy of a BackendQueue instance when needed:

```
public func copy() -> BackendQueue<T> {
    return BackendQueue<T>(items)
}
```

It would be very easy to make the private initializer of the BackendQueue type public, and then allow the main queue type that uses the BackendQueue to call that initializer to create a copy of itself. However, it is generally considered good practice to keep the logic needed to create the copy within the BackendQueue type itself. The reason for this is that if you ever need to make changes to the BackendQueue type, those changes may affect how the type is copied. By having the copy logic embedded within the BackendQueue type itself, it becomes easier to find and modify that logic as needed. Additionally, if you use the BackendQueue type as the backend storage for multiple queue types, this approach ensures that you only need to make changes to the copy logic in one place and also reduces duplicate code. This helps maintain consistency and simplifies the overall code base.

Here is the final code for the BackendQueue type:

```
fileprivate class BackendQueue<T> {
    private var items = [T]()
    public init() {}
    private init(_ items: [T]) {
```

```
            self.items = items
        }
        public func addItem(item: T) {
            items.append(item)
        }
        public func getItem() -> T? {
            if items.count > 0 {
                return items.remove(at: 0)
            } else {
                return nil
            }
        }
        public func count() -> Int {
            return items.count
        }
        public func copy() -> BackendQueue<T> {
            return BackendQueue<T>(items)
        }
    }
}
```

Now let's create our Queue type, which will use the BackendQueue type to implement the copy-on-write feature.

Creating a Queue type

The following code adds the basic queue functionality to our Queue type:

```
struct Queue {
    private var internalQueue = BackendQueue<Int>()
    public mutating func addItem(item: Int) {
        internalQueue.addItem(item: item)
    }
    public mutating func getItem() -> Int? {
        return internalQueue.getItem()
    }
    public func count() -> Int {
        return internalQueue.count()
    }
}
```

The Queue type is implemented as a value type. This type has one private property of the BackendQueue type, which will be used to store the data. This type contains three methods to add items to the queue, retrieve an item from the queue, and return the number of items in the queue.

Now let's explore how we can add the copy-on-write functionality to the Queue type.

Swift has a global function named isKnownUniquelyReferenced(). This function will return true if there is only one reference to an instance of a reference type, or false if there is more than one reference.

We will begin by adding a function to check whether there is a unique reference to the internalQueue instance. This will be a private function named checkUniquelyReferencedInte rnalQueue. The following code shows how we can implement this method:

```
mutating private func checkUniquelyReferencedInternalQueue() {
    if !isKnownUniquelyReferenced(&internalQueue) {
        internalQueue = internalQueue.copy()
        print("Making a copy of internalQueue")
    } else {
        print("Not making a copy of internalQueue")
    }
}
```

In this method, we check to see whether there are multiple references to the internalQueue instances. If there are multiple references, we know that we have multiple copies of the Queue instance and, therefore, we need to create a new copy.

The Queue type itself is a value type; therefore, when we pass an instance of the Queue type within our code, we pass a copy of the instance and not a reference to the instance. The BackendQueue type, which the Queue type uses, is a reference type. Therefore, when a copy is made of a Queue instance, the new copy receives a reference to the original queue's BackendQueue instance and not a new copy. This means that each instance of the Queue type has a reference to the same internalQueue instance. Consider the following code as an example. Both queue1 and queue2 have references to the same internalQueue instance; however, they are different instances:

```
var queue1 = Queue()
var queue2 = queue1
```

In the Queue type, we know that both the addItem() and getItem() methods change the internalQueue instance. Therefore, before we make these changes, we will want to call the checkUniquelyReferencedInternalQueue() method to create a new copy of the internalQueue instance. These two methods will now have the following code:

```
public mutating func addItem(item: Int) {
    checkUniquelyReferencedInternalQueue()
    internalQueue.addItem(item: item)
}
public mutating func getItem() -> Int? {
    checkUniquelyReferencedInternalQueue()
    return internalQueue.getItem()
}
```

With this code, when either the addItem() or getItem() methods are called, which will change the data in the internalQueue instance, we use the checkUniquelyReferencedInternalQueue() method to create a new instance of the data structure.

Let's add one additional method to the Queue type, which will allow us to see whether there is a unique reference to the internalQueue instance or not. Here is the code for this method:

```
mutating public func uniquelyReferenced() -> Bool {
    return isKnownUniquelyReferenced(&internalQueue)
}
```

Here is the full code listing for the Queue type:

```
struct Queue {
    private var internalQueue = BackendQueue<Int>()
    mutating private func checkUniquelyReferencedInternalQueue() {
        if !isKnownUniquelyReferenced(&internalQueue) {
            print("Making a copy of internalQueue")
            internalQueue = internalQueue.copy()
        } else {
            print("Not making a copy of internalQueue")
        }
    }
    public mutating func addItem(item: Int) {
        checkUniquelyReferencedInternalQueue()
        internalQueue.addItem(item: item)
```

```
    }
    public mutating func getItem() -> Int? {
        checkUniquelyReferencedInternalQueue();
        return internalQueue.getItem()
    }
    public func count() -> Int {
        return internalQueue.count()
    }
    mutating public func uniquelyReferenced() -> Bool {
        return isKnownUniquelyReferenced(&internalQueue)
    }
}
```

Now, let's examine how the copy-on-write functionality works with the Queue type.

Using the Queue type

We will start off by creating a new instance of the Queue type, adding an item to the queue, and then checking whether we have a unique reference to the internalQueue instance. The following code demonstrates how to do this:

```
var queue3 = Queue()
queue3.addItem(item: 1)
print(queue3.uniquelyReferenced())
```

When we add the item to the queue, the following message is printed to the console. This message tells us that within the checkUniquelyReferencedInternalQueue() method, it was determined that there was only one reference to the internalQueue instance:

```
Not making a copy of internalQueue
```

Now let's make a copy of the queue3 instance by passing it to a new variable, as follows:

```
var queue4 = queue3
```

Now let's check whether we have a unique reference to the internalQueue instances of either the queue3 or queue4 instance. The following code will do this:

```
print(queue3.uniquelyReferenced())
print(queue4.uniquelyReferenced())
```

This code will print two `false` messages to the console, letting us know that neither instance has a unique reference to its `internalQueue` instances. Now let's add an item to either one of the queues. The following code will add another item to the `queue3` instance:

```
queue3.addItem(item: 2)
```

When we add the item to the queue, we will see the following message printed to the console:

```
Making a copy of internalQueue
```

This message tells us that when we add a new item to the queue, a new copy of the `internalQueue` instance is created. In order to verify this, we can print the results of the `uniquelyReferenced()` methods to the console again. If you check this, you will see that two `true` messages are printed to the console rather than two `false` messages. We can now add additional items to the queues, and we will see that we are no longer creating new instances of the `internalQueue` instance because each instance of the `Queue` type now has its own copy `internalQueue` instance.

If you are planning on creating your own data structure that may contain a large number of items, it is recommended that you implement it with the copy-on-write feature as described here.

Summary

In this chapter, we discussed the differences between value types and reference types. We also explored how reference types can be used to create recursive data structures. Furthermore, we looked into inheritance for reference types and how dynamic dispatch functions. Additionally, we examined how to implement copy-on-write for value types that might contain large data structures.

Now that we have looked at the differences between value and reference types, in the next chapter, let's look at how powerful one particular value type is. That value type is the enumeration.

Unlock this book's exclusive benefits now

Scan this QR code or go to packtpub.com/unlock, then search this book by name.

Note: Keep your purchase invoice ready before you start.

6

Enumerations

In most modern programming languages, enumerations serve as essential tools for grouping a set of related values and improving the readability, organization, and maintainability of code. However, not all enumerations are created equal. Swift, with its modern and innovative approach to language design, has taken enumerations to a whole new level, offering a wide range of features and capabilities that set them apart from enumerations in other languages.

Swift enumerations may appear modest at first glance, but they possess a depth and versatility that far exceed the traditional concepts of enumerations found in other languages. In this chapter, we will look into these powerful features to help you fully unlock their potential.

In this chapter, we will cover the following topics:

- How to use raw and associated values with enumerations
- How to use pattern matching with enumerations
- How to iterate over enumeration cases
- How to use methods and properties with enumerations

Let's look at some of the exciting features that enumerations in Swift offer, starting with raw values.

Raw values

Enumerations in Swift can be defined as having raw values, offering a way to associate constant values with their cases. These raw values can be strings, characters, or any of the integer or floating-point number types. By assigning raw values to enumeration cases, the enumeration can become easier to use with existing code that expects those types. Enumerations can also be initialized using their raw value. Raw values can be particularly useful when we need to map enumeration cases to specific data.

In order to define an enumeration with raw values, we simply need to specify the raw value type that the enumeration should use. Let's look at a basic example:

```
enum Direction: String {
    case north = "N"
    case south = "S"
    case west = "W"
    case east = "E"
}
```

In this example, we are defining the four points of a compass. In the enumeration declaration, we are defining that the raw value type is going to be a String. With each case, we are also defining the raw values. We could access the raw values like this:

```
let myDirection = Direction.north
print("My direction is: \(myDirection.rawValue)")
```

In this code, we set the `myDirection` constant to the `Direction` case of North. We then access the raw value using the `rawValue` property. The output of this code is the following:

```
My direction is: N
```

When we create an enumeration with a raw value defined as an integer, we do not need to define the raw values for each of the cases. For example, let's look at the following enumeration:

```
enum Month: Int {
    case January = 1, February, March, April, May, June, July, August,
        September, October, November, December
}
```

In this example, we define a `Month` enumeration, which has an integer as its raw value type. We set a raw value of 1 for the January case. By establishing the enumeration with an integer raw value and initializing January with a value of 1, subsequent months will automatically increment by one. Therefore, February will automatically be assigned a raw value of 2, March will have 3, and this pattern will continue accordingly for each subsequent month.

When using an integer as the raw value, if no initial value is set, the raw value will automatically start with 0.

One of the benefits of raw values is that Swift automatically provides an initializer for enumerations with raw values. This initializer takes a parameter of the raw value's type and returns an
optional with either an enumeration case or `nil` if there's no matching case for the raw value.
This code illustrates how this works:

```
if let month = Month(rawValue: 4) {
    print("The month is \(month)")
}
```

In this example, we set the `month` constant to the `Month` case that has the raw value of 4. The
output of this code is the following:

```
The month is April
```

One use case for using raw values with enumerations is storing a collection of constants instead
of creating a static type. For example, if we were building an application that required multiple
URLs, we could create an enumeration to store those URLs like this:

```
enum UrlList: String {
    case myUrl = "http://myurl.com"
    case serviceUrl = "http://serviceurl.com"
}
```

We could then use the URLs by using their raw values, as shown here:

```
let serviceUrl = URL(string: UrlList.serviceUrl.rawValue)
```

Now that we have seen how to use raw values with enumerations, let's look at associated values.

Associated values

In Swift, enumerations can include associated values, which enable us to store additional data
alongside the enumeration cases. This additional information can be of any type and vary between
the different cases within the same enumeration.

Let's look at how we might use associated values by defining a `Product` enumeration that contains two product types:

```
enum Product {
    case book(Double, Int, Int)
    case puzzle(Double, Int)
}
```

In this code, we are defining a `Product` enumeration with two members: book and puzzle. The book member has associated values of the `Double`, `Int`, and `Int` types, while the `puzzle` member has associated values of the `Double` and `Int` types.

Let's create a function that will randomly generate either a new book or a new puzzle. We will create the function like this:

```
func getProduct() -> Product {
    if Bool.random() {
        return Product.book(49.99, 2025 394)
    } else {
        return Product.puzzle(9.99, 200)
    }
}
```

In this function, we use the `random()` method of the Boolean type. If it comes back as true, we create a book, and if it comes back as false, we create a puzzle.

```
var product = getProduct()

switch product {
case .book(let price, let year, let pages):
    print("Product was published in \(year) for the price of \(price) and
            has \(pages) pages")
case .puzzle(let price, let pieces):
    print("Product is a puzzle with \(pieces) pieces and sells for
            \(price)")
}
```

In this example, we use the `getProduct()` function to randomly generate a book or a puzzle and then use the `switch` statement to determine which type was created. The `switch` statement is structured in a way that allows it to execute different blocks of code based on which case is assigned. The associated values are extracted within the `switch` statement. In this example, we extracted the associated values as constants using the `let` keyword.

When this code is executed, we should see one of the two outputs shown:

```
Mastering Swift was published in 2024 for the price of 49.99 and has 394 pages
World puzzle is a puzzle with 200 pieces and sells for 9.99
```

Associated values are incredibly useful for representing cases where each enumeration case can carry additional data or information. Having the ability to vary the data by case makes associated values very flexible.

Another useful feature of associated values is the ability to use labels, which makes the code easier to read and understand, especially when working with complex data. To use labels, we could rewrite our Product enumeration like this:

```
enum ProductWithLabels {
    case book(price: Double, yearPublished: Int, pageCount: Int)
    case puzzle(price: Double, pieceCount: Int)
}
```

We could then use this new enumeration, as shown here:

```
let masterSwift = ProductWithLabels.book(price: 49.99,
    yearPublished: 2024, pageCount: 394)
```

As we can see, this makes the code easier to read and understand.

When we used the `switch` statement with enumerations, we were showing how to use pattern matching. Let's look at this further in the next section, and while we are at it, let's look at another example of using associated values.

Pattern matching

Pattern matching is a powerful feature in Swift that enables us to compare values against a set of patterns and execute corresponding code blocks. As we saw in the previous example, pattern matching is very well suited for working with enumerations. As another example of pattern matching, let's create a new enumeration:

```
enum Weather {
    case sunny
    case cloudy
    case rainy(Int)
    case snowy(amount: Int)
}
```

This example shows how truly flexible associated values can be with enumerations. In this example, we create a `weather` enumeration with four cases; however, only two of the cases have associated values, and the snowy case has a named associated type. Now, let's see how we can use the `switch` statement to match the different cases by creating a `showWeather()` function:

```
func showWeather(_ weather: Weather) {

    switch weather {
    case .sunny:
        print("It's sunny")
    case .cloudy:
        print("It's cloudy")
    case .rainy(let intensity):
        print("It's raining with an intensity of \(intensity).")
    case .snowy(let amount):
        print("It's snowing with an estimated amount of \(amount).")
    }

}
```

This example looks very similar to the one that we used earlier for the `Product` enumeration, but notice that we are only getting the associated values for the cases that have them. With Swift pattern matching, we can also match on multiple cases; the following code shows how to do this:

```
func showPrecipitation(_ weather: Weather) {
    switch weather {
    case .sunny, .cloudy:
        print("No precipitation today")
    case .rainy(let intensity):
        print("It rained with an intensity of \(intensity).")
    case .snowy(let amount):
        print("It snowed \(amount) inches.")
    }
}
```

In this `switch` statement, the first case matches both the `.sunny` and `.cloudy` cases for the `Weather` enumeration, indicating that there was no precipitation for that day. The next two cases match the `.rainy` and `.snowy` cases and print out messages specific to each case. We would use the `Weather` enumeration and these two functions like this:

```
showWeather(.sunny)
showWeather(.rainy(2))
showWeather(.snowy(amount: 8))

showPrecipitation(.sunny)
showPrecipitation(.rainy(2))
showPrecipitation(.snowy(amount: 8))
```

If this code was run, the following output would be displayed:

```
It's sunny
It's raining with an intensity of 2.
It's snowing with an estimated amount of 8.
No precipitation today
It rained with an intensity of 2.
It snowed 8 inches.
```

We are not limited to only using pattern matching with the switch statement. As an example, we could also use it with an if statement using the if case syntax. Let's look at how this works with the following code:

```
func tooWet(_ weather: Weather) {
    if case let .rainy(intensity) = weather, intensity > 5 {
        print("Too wet to go outside")
    }
}
```

In this code, we use the if case syntax to see if the intensity of the rain is greater than 5 and if so, we let the user know it is too wet to go outside.

> Pattern matching with enumerations can be used with the switch, if case, and guard statements as well as the for case and while loops.

Now, let's look at how we would iterate over enumerations.

Enumeration iteration

Iterating over enumeration cases is a common task in Swift development, as it enables you to perform various operations on the entire set of values. This can be particularly useful when you wish to perform an action for each case in the enumeration.

Let's look at how we can iterate over an enumeration. We will start by defining an enumeration to use:

```
enum DaysOfWeek: String, CaseIterable {
    case Monday = "Mon"
    case Tuesday = "Tues"
    case Wednesday = "Wed"
    case Thursday = "Thur"
    case Friday = "Fri"
    case Saturday = "Sat"
    case Sunday = "Sun"
}
```

This enumeration looks very similar to the enumerations that we defined in the *Raw values* section of this chapter, and it has a raw value type defined as a String type. However, in this example, we are specifying that the DaysOfWeek enumeration conforms to the CaseIterable protocol. This will enable us to iterate over the different cases of the enumeration. Let's see how we would do this:

```
for day in DaysOfWeek.allCases {
    print("    -- \(day)")
}
```

The CaseIterable protocol enables us to access all of the enumeration cases using the allCases property. The for-in loop is then used to iterate over each case. The following shows the output of this code:

```
    -- Monday
    -- Tuesday
    -- Wednesday
    -- Thursday
    -- Friday
    -- Saturday
    -- Sunday
```

There are times when we might wish to iterate over a subset of the enumeration cases. While Swift does not provide built-in functionality for this, we can enable it by filtering the `allCases` property, as shown here:

```
for weekDay in DaysOfWeek.allCases.filter(
    { $0 != .Saturday && $0 != .Sunday }) {
        print("   -- \(weekDay).")
}
```

In this example, we are filtering out the weekend days and only looping through the weekdays. The output would look like this:

```
   -- Monday.
   -- Tuesday.
   -- Wednesday.
   -- Thursday.
   -- Friday.
```

Another thing we can do as we loop through an enumeration is to get the index of the case. This next example illustrates this:

```
for (index, day) in DaysOfWeek.allCases.enumerated() {
    print("  --\(index): \(day)")
}
```

In this example, we use the `enumerated()` method, which returns a sequence of tuples containing the index and the case value. The output would look like this:

```
  --0: Monday
  --1: Tuesday
  --2: Wednesday
  --3: Thursday
  --4: Friday
  --5: Saturday
  --6: Sunday
```

As we iterate through the cases of an enumeration, we can access the raw values of the cases as well:

```
for day in DaysOfWeek.allCases {
    print("   -- \(day.rawValue): \(day) ")
}
```

This code prints out both the raw value and the case. The output would look like this:

```
-- Mon: Monday
-- Tues: Tuesday
-- Wed: Wednesday
-- Thur: Thursday
-- Fri: Friday
-- Sat: Saturday
-- Sun: Sunday
```

So far in this chapter, we have seen how Swift's enumerations are more powerful than the other programming languages and have features that enumerations in other languages do not have. Enumerations in Swift can actually also be used in ways similar to structures and classes, as we will see in the next section.

Going beyond basic values

While enumerations are typically used to define a set of related values, they can also be used similarly to how we might use a structure or class. This enables us to add additional functionality and state within our enumeration cases, making them even more versatile and powerful.

To see how enumerations are similar to structures and classes, let's begin by seeing how we would add a computed property and methods to an enumeration. The following example creates an enumeration with a computed property method:

```swift
enum Priority {
    case low, medium, high, critical

    var isHigh: Bool {
        self == .high || self == .critical
    }

    func description() -> String {
        switch self {
        case .low:
            "Low Priority"
        case.medium:
            "Medium Priority"
        case .high:
            "High Priority"
```

```
            case .critical:
                "Critical Priority"
            }
        }
    }
```

In this example, we create an enumeration named `Priority` with four cases and add a parameter named `isHigh` of the Boolean type and a `description()` method, which returns a string type describing the priority level. This enumeration can now be used with its properties and methods, as shown in the following code:

```
let priority = Priority.high
print("This is a \(priority.description()) and needs to be done now
    \(priority.isHigh)")
```

As seen in this example, computed properties and methods can be incorporated into an enumeration, similar to how we would include them in a structure or class, and they are utilized in a similar manner as well.

Another feature of Swift enumeration is that they can adopt protocols. To see this, let's begin by creating a protocol for our types to use:

```
protocol Describable {
    func description() -> String
}
```

The `Describable` protocol defines one method that any type that conforms to it must implement, which is the `description()` method. We could create an enumeration that conforms to this protocol like this:

```
enum TrafficLight: Describable {
    case red
    case yellow
    case green

    func description() -> String {
        switch self {
        case .red:
            return "Stop"
        case .yellow:
```

```
            return "Proceed with caution"
        case .green:
            return "Go"
        }
    }
}
```

In the `TrafficLight` enumeration, we specify that it will conform to the `Describable` protocol, and the `description()` method was added to make it conform. What makes this so powerful is if we define that the `Priority` enumeration also conforms to the `Describable` protocol like this:

```
enum Priority: Describable {

    ...

}
```

We can then use both enumerations together like this:

```
let describe:[any Describable] = [TrafficLight.green, Priority.high]

for item in describe {
    print("Description: \(item.description())")
}
```

In this code, we define an array of items that conform to the `Describable` protocol and then iterate through that array, showing the descriptions.

Let's take this example a step further and create a structure that also conforms to the `Describable` protocol like this:

```
struct Person: Describable {
    var firstName: String
    var lastName: String

    func description() -> String {
        "\(firstName) \(lastName)"
    }
}
```

We can now use instances of the enumerations interchangeably with instances of the Person structure, as shown here:

```
let describe:[any Describable] = [TrafficLight.green, Priority.high,
person(firstName: "Jon", lastName: "Hoffman")]

for item in describe {
    print("Description: \(item.description())")
}
```

In this section, we have shown how enumerations are much more similar to fully fledged types in Swift as compared to other languages. They can contain methods and computed properties as well as adopting protocols.

One thing to keep in mind with Swift enumerations is they are a value type similar to structures. Therefore, when instances of the enumeration are passed within your code, they are passed by value and not reference.

Summary

In this chapter, we looked at some of the powerful capabilities of Swift enumerations. We saw how the support for raw values offers greater flexibility with a wide range of types, giving enumerations greater adaptability compared to other languages' limited raw value options.

We saw how enumerations can have associated values, allowing for flexible representation of complex data structures, a feature that distinguishes them from enumerations in other languages. We also looked at how pattern matching in Swift leads to more concise and readable code.

We also covered how Swift enumerations provide native support for methods and computed properties, making them similar to structures and classes. Swift's value semantics ensure safer concurrency and immutability.

In the next chapter, we'll look at reflection and how that is enabled in Swift through the Mirror API.

Unlock this book's exclusive benefits now

Scan this QR code or go to packtpub.com/unlock, then
search this book by name.

Note: Keep your purchase invoice ready before you start.

7

Reflection

Reflection is a powerful feature that enables a program to inspect and, with some languages, modify its own structure and behavior at runtime. This means that a program can analyze and modify its variables, data types, properties, and methods dynamically, without needing to know about them at compile time. Reflection is used for a variety of purposes, such as debugging, flexible frameworks, and performing dynamic analysis or modifications of program components.

The Mirror API enables reflection in Swift. This is a component of the Swift standard library that provides a means for examining the properties, types, and values of instances at runtime. Unlike some other languages that enable modification of objects at runtime, Swift's reflection capabilities are designed with a focus on examining rather than modifying. This design choice aligns with Swift's emphasis on type safety and performance.

In this chapter, we will cover the following topics:

- What reflection is and why it is important
- How to use the Mirror API to inspect an instance of a type
- How to use the Mirror API to inspect the properties of an instance
- How to use the Mirror API to serialize an instance

Reflection and the Mirror API

Reflection allows applications to inspect, analyze, and modify their structure and behavior at runtime. With reflection, we have the ability to discover information about types at runtime. We can think of reflection as a way for a program to look at and inspect itself.

Reflection in Swift is implemented using the Mirror API. This API enables us to inspect instances of classes, structs, and enumerations, offering valuable insights into the instance's type and the values of its properties. This capability proves especially beneficial when handling dynamic or unknown types, as it enables the querying of an instance without prior knowledge of its exact type.

Through the Mirror API, we can access the instance's properties, termed as children, providing a means to examine the stored values within an instance. This is particularly useful when dealing with complex data structures or when seeking to comprehend the internal mechanisms of a specific type. Let's explore what the Mirror API can do.

Before deciding to use Swift's reflection capabilities, we must take into account the performance implications. Reflective tasks typically require more computing power than direct access methods. This increased overhead means that using reflection extensively, in parts of the application that are critical for performance, is not recommended.

The Mirror API gives us access to information about Swift objects at runtime. Let's take a look at this.

Getting started with the Mirror API

Before we explore the capabilities of the Mirror API, we will need to create a type that can be used for the examples. Let's create the following Person structure:

```
struct Person {
    var firstName: String
    var lastName: String
    var age: Int
}
```

In this structure, we define three properties, firstName and lastName (both of which are strings), and age, which is an integer. Now, let's see how we would use the Mirror API with our Person type:

```
let person = Person(firstName: "Jon", lastName: "Hoffman", age: 55)
let mirror = Mirror(reflecting: person)
```

This code starts off by creating an instance of the Person type. We then create an instance of the Mirror type using the Person instance. We can now use the mirror instance to access different attributes of the person instance.

Let's look at the displayStyle and subjectType values of our `person` instance using the mirror instance. The following code will display them:

```
print("Display Style: \(mirror.displayStyle!)")
print("Subject Type: \(mirror.subjectType)")
```

> displayStyle is an enumeration that contains the following cases:
>
> ```
> case class
> case collection
> case dictionary
> case enum
> case optional
> case set
> case struct
> case tuple
> ```
>
> From this list, we can see that displayStyle tells us the underlying type of the instance.
>
> subjectType is the name of the type.

If we ran this code, we would see the following output:

```
Display Style: Optional(Swift.Mirror.DisplayStyle.struct)
Subject Type: Person
```

The Mirror API could also be used to inspect the properties of the instance. Let's see how we could do this:

```
for (label, value) in mirror.children {
    print("Property: \(label ?? "Unknown"), Value: \(value)")
}
```

The children property contains a list of the instance's properties and their values. It's important to note that this property only contains the stored properties of the instance, and other elements, such as computed properties, are not part of this collection. The output of this code would look something like this:

```
Property: firstName, Value: Jon
Property: lastName, Value: Hoffman
Property: age, Value: 55
```

Let's look at another example, but this time, we will use the following class hierarchy:

```
class Vehicle {
    var numberOfWheels:    Int

    init(numberOfWheels: Int) {
        self.numberOfWheels = numberOfWheels
    }
}

class Car: Vehicle {
    var numberOfDoors:    Int

    init(numberOfDoors: Int) {
        self.numberOfDoors = numberOfDoors
        super.init(numberOfWheels: 4)
    }
}
```

In this code, we have a Vehicle superclass with a Car subclass that inherits from the Vehicle class. Now, let's create an instance of the Car type and look at the DisplayStyle and subjectType properties using the following code:

```
let car = Car(numberOfDoors: 4)
let mirrorCar = Mirror(reflecting: car)

print("Display Style: \(mirrorCar.displayStyle!)")
print("Subject Type: \(mirrorCar.subjectType)")
```

This code will display what we probably expect, which is the following:

```
Display Style: Optional(Swift.Mirror.DisplayStyle.class)
Subject Type: Car
```

We could also display the properties of the car instance using the following code:

```
for (label, value) in mirrorCar.children {
    print("Property: \(label ?? "Unknown"), Value: \(value)")
}
```

This is where things are a little different when it comes to class hierarchy and the Mirror API, because the preceding code would have the following output:

```
Property: numberOfDoors, Value: 4
```

You will notice that only the properties of the Car type are displayed, and the properties of the class's superclass are not. In order to inspect the properties defined in the superclass, we would need to use superClassMirror to retrieve an instance of the Mirror API for the superclass. The following code shows how we would do this:

```
if let mirrorCarSuper = mirrorCar.superclassMirror {
    print("Super Class: \(mirrorCarSuper.subjectType)")

    for (label, value) in mirrorCarSuper.children {
        print("Property: \(label ?? "Unknown"), Value: \(value)")
    }
}
```

This code will look to see if there is a superclass, and if there is, output the subjectType and properties of the superclass. The output of this code would look like this:

```
Super Class: Vehicle
Property: numberOfWheels, Value: 4
```

While the Mirror API's default reflection capabilities provide a basic representation of any type, there are cases where we need more control over how our types are represented. For this, we can use the CustomReflectable protocol.

CustomReflectable protocol

By conforming to the CustomReflectable protocol, we can define exactly what our types present when they are inspected at runtime. This is particularly useful for debugging and logging, and it also allows us to hide sensitive properties for security purposes.

To conform to the CustomReflectable protocol, we must implement a single property of the Mirror type. Creating a custom Mirror instance involves defining the subject, which is generally the instance itself; the children, which is a collection of key-value pairs representing the properties; and the displayStyle, which we saw previously in this chapter.

Let's explore how we would use this protocol by first defining a new Person type that looks like this:

```
struct Person {
    var firstName: String
    var lastName: String
    var userName: String
    var age: Int
    var password: String
}
```

This Person type simply defines five properties that make up a person. Now, let's extend this person type and have it conform to the CustomReflectable protocol:

```
extension Person: CustomReflectable {
    var customMirror: Mirror{
        let fullName = "\(firstName) \(lastName)"
        return Mirror(self, children: [
            "firstName": firstName,
            "lastName": lastName,
            "userName": userName,
            "fullName": fullName,
            "age": String(age)
        ], displayStyle: .struct)
    }
}
```

This code extends the Person type to conform to the `CustomReflectable` protocol by implementing the `customMirror` property. Within this property, an instance of the `Mirror` type is created, which begins with a new computed property named `fullName`; this combines the `firstName` and `lastName` properties. The Mirror instance is initialized with the current instance (self) as the subject, a collection of key-value pairs representing the properties we wish to reflect, and then the display style.

A couple of important details within the key-value pairs to note are that we are able to reflect the `fullName` property as if it was a standard property of the Person type, and we also excluded the `password` property, preventing it from being reflected for security reasons.

We could use the Person type and have it reflected like this:

```
let person = Person(firstName: "Jon", lastName: "Hoffman", userName:
"MyUser", age: 55, password: "MyPass123!")
let mirror = Mirror(reflecting: person)

for (label, value) in mirror.children {
    print("Property: \(label ?? "Unknown"), Value: \(value)")
}

print("Display Style: \(mirror.displayStyle!)")
print("Subject Type: \(mirror.subjectType)")
```

When we run this code, we will see the following output:

```
Property: firstName, Value: Jon
Property: lastName, Value: Hoffman
Property: userName, Value: MyUser
Property: fullName, Value: Jon Hoffman
Property: age, Value: 55
Display Style: struct
Subject Type: Person
```

Another thing to note is that we have the ability to change the display style to any valid value, and it will reflect the value that we define and not the actual type. Therefore, if we defined the display style as .class, it would reflect that our type is a class and not a structure.

Now that we have seen how to use the Mirror API, let's look at how we can use it to serialize objects.

Serializing objects

While the Mirror API isn't designed for serialization like Codable or JSONSerialization, it can be used to manually serialize an object by examining its properties. The following code illustrates how to do this:

```
func serialize<T>(_ value: T) -> [String: Any] {
    let mirror = Mirror(reflecting: value)
    var result = [String: Any]()

    for child in mirror.children {
        if let propertyName = child.label {
```

```
                    result[propertyName] = child.value
            }
        }

    return result
}
```

In this code, we create a generic function called `serialize` that accepts an instance of any type and returns a dictionary instance. Within the instance, we create an instance of the Mirror API and then loop through the properties, adding the property name and value to the dictionary that we will return. We could use this code like this:

```
let serializedPerson = serialize(person)
print(serializedPerson)
```

If we used the same person instance that we created earlier in the chapter, this code would have the following output:

```
["firstName": "Jon", "lastName": "Hoffman", "age": 55]
```

This approach will work for simple, flat data structures; the Mirror API is not designed to serialize complex data structures or nested structures. For those, we would want to use the `Codable` protocol, which is written for encoding and decoding complex data structures and provides more robust and comprehensive serialization capabilities.

Using the Mirror API like this is primarily useful for debugging or similar use cases where traditional serialization tools are not sufficient. For production-level code involving JSON data, it would be best to use the `Codable` protocol with `JSONEncoder` and `JSONDecoder`.

Also, keep in mind what we mentioned earlier about the performance implications of using reflection – we should be careful not to overuse it, especially in performance-critical parts of our application, because it will adversely affect performance.

Summary

In Swift, the Mirror API enables reflection, making it possible to inspect the properties, types, and values of instances at runtime.

In this chapter, we showed how we could use the Mirror API to inspect the properties of a structure and class, revealing details such as property names and values, and even the instance's display style and underlying type category. Advanced use cases for the Mirror API include debugging complex data structures, custom serialization, and dynamically exploring class hierarchies, including superclass properties. However, while the Mirror API can be used for serialization by manually iterating over properties, it is not designed for handling complex or nested data structures as robustly as the `Codable` protocol.

There are also performance implications to consider, as reflection requires more computing resources than direct access methods and is not recommended for performance-critical parts of an application.

In the next chapter, we will look at the Swift availability feature and error handling.

Unlock this book's exclusive benefits now

Scan this QR code or go to `packtpub.com/unlock`, then search this book by name.

Note: Keep your purchase invoice ready before you start.

8

Error Handling and Availability

Managing unforeseen issues and ensuring that applications are always available are critical components of high-quality software. This chapter will examine how Swift provides error handling, which enables developers to anticipate and manage potential failures in a clean and efficient manner.

Ensuring that our applications utilize the most up-to-date and supported features while still providing backward compatibility is essential. Using the availability attribute effectively enables developers to gracefully manage the introduction of new features and deprecations, while ensuring that applications remain accessible and functional across an array of devices and Swift versions.

Together, mastering error handling and the availability attribute equips Swift developers with the tools needed to build resilient and adaptable applications.

We will cover the following topics in this chapter:

- How to represent errors
- How to use the do-catch block in Swift
- How to use the defer statement
- How to use typed throws
- How to use the availability attribute

Let's get started by looking at how Swift provides error handling.

Native error handling

Languages such as Java and C# generally refer to the error-handling process as exception handling. Within the Swift documentation, Apple refers to this process as error handling. While, on the outside, Java and C# exception handling may look similar to Swift's error handling, there are some significant differences that those familiar with exception handling in other languages will notice throughout this chapter.

Representing errors

Before we can really understand how error handling works in Swift, we must see how we would represent an error. In Swift, errors are represented by values of types that conform to the `Error` protocol. Swift's enumerations are very well suited to modeling error conditions because we generally have a finite number of error conditions to represent.

Let's look at how we would use an enumeration to represent an error. For this example, we will define an error named `MyError` with three error conditions – `minor`, `bad`, and `terrible`:

```
enum MyError: Error {
    case minor
    case bad
    case terrible
}
```

In this example, we defined that the `MyError` enumeration conforms to the `Error` protocol and defined three error conditions. That is all there is to defining basic error conditions.

> 💡 **Quick tip**: Enhance your coding experience with the **AI Code Explainer** and **Quick Copy** features. Open this book in the next-gen Packt Reader. Click the **Copy** button
>
> **(1)** to quickly copy code into your coding environment, or click the **Explain** button
>
> **(2)** to get the AI assistant to explain a block of code to you.

```
                                                          Copy      Explain
function calculate(a, b) {
  return {sum: a + b};                                     1          2
};
```

> 📖 **The next-gen Packt Reader** is included for free with the purchase of this book. Scan the QR code OR go to `packtpub.com/unlock`, then use the search bar to find this book by name. Double-check the edition shown to make sure you get the right one.

We can also use the associated values with our error conditions to add more details. Let's say that we want to add a description to the `terrible` error condition. We would do so like this:

```
enum MyError: Error {
    case minor
    case bad
    case terrible(description:String)
}
```

Those who are familiar with exception handling in Java and C# can see that representing errors in Swift is a lot cleaner and easier because we do not need to create a lot of boilerplate code or a full class. With Swift, it can be as simple as defining an enumeration with our error conditions. Another advantage is that it is very easy to define multiple error conditions and group them together so that all the related error conditions are included in one type.

Now, let's learn how to model errors in Swift. For this example, we'll look at how we would assign numbers to players on a baseball team. On a baseball team, every new player who is called up is assigned a unique number. This number must also be within a certain range (0–99) because a player's jersey is only big enough for two numerals.

For this example, we would have three error conditions: the number is too large, the number is too small, and the number is not unique. The following example shows how we might represent these error conditions:

```
enum PlayerNumberError: Error {
    case numberTooHigh(description: String)
    case numberTooLow(description: String)
    case numberAlreadyAssigned
}
```

With the `PlayerNumberError` type, we define three very specific error conditions that tell us exactly what went wrong when we tried to assign a number to a player. These error conditions are also grouped together in one type because they are all related to assigning numbers to a player.

Now that we know how to represent errors, let's look at how to throw errors.

Throwing errors

When an error occurs in a function, the code that called the function must be made aware of it; this is called **throwing an error**. When a function throws an error, it assumes that the code that called the function, or some code further up the chain, will catch and recover appropriately from the error.

To throw an error from a function, we use the throws keyword. This keyword lets the code that called it know that an error may be thrown from the function.

Before we see how to throw an error, let's add a fourth error to the PlayerNumberError type that we defined earlier. This demonstrates how easy it is to add error conditions to our error types. This error condition, named numberDoesNotExist, is thrown if we are trying to retrieve a player by their number, but no player has been assigned that number. The new PlayerNumberError type will now look similar to this:

```
enum PlayerNumberError: Error {
    case numberTooHigh(description: String)
    case numberTooLow(description: String)
    case numberAlreadyAssigned
    case numberDoesNotExist
}
```

To demonstrate how to throw errors, let's create a BaseballTeam structure that will contain a list of players for a given team. These players will be stored in a dictionary object named players. We will use the player's number as the key because we know that each player must have a unique number. The BaseballPlayer type, which will be used to represent a single player, will be a type alias for a tuple type, and is defined like this:

```
typealias BaseballPlayer = (firstName: String, lastName: String,
                            number: Int)
```

Let's start by defining the basics of the BaseballTeam structure. The following code creates the structure and defines three private properties that we will use within the structure:

```
struct BaseballTeam {
    private let maxNumber = 99
    private let minNumber = 0
    private var players = [Int: BaseballPlayer]()


}
```

The `maxNumber` and `minNumber` properties contain the maximum and minimum numbers that a player can have, and the `players` dictionary contains a list of players on the team.

In this `BaseballTeam` structure, we will add two methods. The first one will be named `addPlayer()`. This method will accept one parameter of the `BaseballPlayer` type and attempt to add the player to the team. This method can also throw one of three error conditions: `numberTooHigh`, `numberTooLow`, or `numberAlreadyExists`. Here is how we would write this method:

```
mutating func addPlayer(player: BaseballPlayer) throws {
    guard player.number < maxNumber else {
        throw PlayerNumberError.numberTooHigh(description:
                        "Max number is \(maxNumber)")
    }
    guard player.number > minNumber else {
        throw PlayerNumberError.numberTooLow(description:
                        "Min number is \(minNumber)")
    }
    guard players[player.number] == nil else {
        throw PlayerNumberError.numberAlreadyAssigned
    }
    players[player.number] = player
}
```

We can see that the `throws` keyword is added to the method's definition. We then use the three guard statements to verify that the number is not too large or too small, and is unique in the `players` dictionary. If any of these conditions are not met, we throw the appropriate error using the throw keyword. If we make it through all three checks, the player is then added to the `players` dictionary.

The second method that we will be adding is the `getPlayerByNumber()` method. This method will attempt to retrieve the baseball player that has been assigned a given number. If no player is assigned that number, this method will throw a `numberDoesNotExist` error. The `getPlayerByNumber()` method will look similar to this:

```
func getPlayerByNumber(number: Int) throws -> BaseballPlayer {
    if let player = players[number] {
        return player
    } else {
        throw PlayerNumberError.numberDoesNotExist
    }
}
```

We have added the throws keyword to this method definition as well; however, this method also has a return type. When we use the throws keyword with a return type, it must be placed before the return type in the method's definition.

Within the method, we attempt to retrieve the baseball player with the number that is passed into the method. If we can retrieve the player, we return it; otherwise, we throw the numberDoesNotExist error. Note that if we throw an error from a method that has a return type, a return value is not required if an error is returned.

Catching errors

In this section, we'll explore how to catch errors with Swift.

Using the do-catch block

When an error is thrown from a function, we need to catch it in the code that called it; this is done using the do-catch block. We use the try keyword within the do-catch block to identify the places in the code that may throw an error. The do-catch block with a try statement has the following syntax:

```
do {
    try [Some function that throws]
    [Code if no error was thrown]
} catch [pattern] {
    [Code if function threw error]
}
```

If an error is thrown, it is propagated out until it is handled by a catch clause. The catch clause consists of the catch keyword, followed by a pattern to match the error against. If the error matches the pattern, the code within the catch block is executed.

Let's look at how to use the do-catch block by calling both the getPlayerByNumber() and addPlayer() methods of the BaseballTeam structure. Let's look at the getPlayerByNumber() method first, since it only throws one error condition:

```
do {
    let player = try myTeam.getPlayerByNumber(number: 34)
    print("Player is \(player.firstName) \(player.lastName)")
} catch PlayerNumberError.numberDoesNotExist {
    print("No player has that number")
}
```

Within this example, the do-catch block calls the getPlayerByNumber() method of the BaseballTeam structure. This method will throw the numberDoesNotExist error condition if no player on the team has been assigned this number; therefore, we attempt to match this error in the catch statement.

Any time an error is thrown within a do-catch block, the remainder of the code within the block is skipped, and the code within the catch block that matches the error is executed. Therefore, in our example, if the numberDoesNotExist error is thrown by the getPlayerByNumber() method, the print statement (where we output the player's name) is never reached.

If a pattern is not included after the catch statement, or if we put in an underscore, the catch statement will match all the error conditions. For example, either one of the following two catch statements will catch all errors:

```
do {
    // our statements
} catch {
    // our error conditions
}

do {
    // our statements
} catch _ {
    // our error conditions
}
```

If we want to capture the error, we can use the let keyword, as shown in the following example:

```
do {
    // our statements
} catch let error {
    print("Error:\(error)")
}
```

Using multiple catch statements

Now, let's look at how to use multiple catch statements in order to catch different error conditions. To do this, we will call the addPlayer() method of the BaseballTeam structure:

```
do {
    try myTeam.addPlayer(player:("David", "Ortiz", 34))
    print("Player Added")
} catch PlayerNumberError.numberTooHigh(let description) {
```

```
    print("Error: \(description)")
} catch PlayerNumberError.numberTooLow(let description) {
    print("Error: \(description)")
} catch PlayerNumberError.numberAlreadyAssigned {
    print("Error: number already assigned")
}
```

In this example, we have three catch statements. Each catch statement has a different pattern to match; therefore, they will each match a different error condition.

As you may recall, the numberTooHigh and numberTooLow error conditions have associated values. To retrieve the associated values, we use the let statement within parentheses, as shown in this example.

It's always a good practice to have your final catch statement be an empty one. This ensures that it will catch any errors that weren't matched by the patterns in the preceding catch statements. Therefore, the previous example should be rewritten like this:

```
do {
    try myTeam.addPlayer(player:("David", "Ortiz", 34))
} catch PlayerNumberError.numberTooHigh(let description) {
    print("Error: \(description)")
} catch PlayerNumberError.numberTooLow(let description) {
    print("Error: \(description)")
} catch PlayerNumberError.numberAlreadyAssigned {
    print("Error: number already assigned")
} catch {
    print("Error: Unknown Error")
}
```

Using error propagation

We can also let the errors propagate out rather than immediately catching them. To do this, we just need to add the throws keyword to the function definition. For instance, in the following example, rather than catching the error, we could let it propagate out to the code that called the function:

```
func myFunc() throws {
    try myTeam.addPlayer(player:("David", "Ortiz", 34))
}
```

Using a forced-try expression

If we are certain that an error will not be thrown, we could call the function using a forced-try expression, which is written as try!. The forced-try expression disables error propagation and wraps the function call in a runtime assertion so no error will be thrown from this call. If an error is thrown, you will get a runtime error, so be very careful when using this expression.

It is highly recommended that you avoid using the forced-try expression in production code since it can cause a runtime error and cause your application to crash.

Using an optional try

When I work with exceptions in languages such as Java and C#, I see a lot of empty catch blocks. This is where we need to catch the exception because one might be thrown; however, we do not want to do anything with it. In Swift, the code would look something like this:

```
do {
    let player = try myTeam.getPlayerByNumber(number: 34)
    print("Player is \(player.firstName) \(player.lastName)")
} catch {}
```

Code like this is one of the things that I dislike about exception handling; however, in Swift, we do not need to write code like this. Instead, we can use the optional try: try?. This attempts to perform an operation that may throw an error and converts it into an optional value; therefore, the result of the operation will be either nil if an error is thrown or the result of the operation if no error is thrown.

Since the results of try? are returned in the form of an optional, we would normally use this with optional binding. We could rewrite the previous example like this:

```
if let player = try? myTeam.getPlayerByNumber(number: 34) {
    print("Player is \(player.firstName) \(player.lastName)")
}
```

As we can see, this makes our code much cleaner and easier to read.

Next, let's take a look at the LocalizedError protocol and how we can use it to add better descriptions to our errors.

The LocalizedError protocol

While the Error protocol enables us to define basic errors, it is sometimes preferable that we provide more contextual information, such as user-facing error descriptions or recovery suggestions. For this, Swift provides the LocalizedError protocol, which extends the Error protocol by adding several optional properties to provide additional information about the error. These properties are:

- errorDescription: A localized description of the error
- failureReason: A localized reason for why the error occurred
- recoverySuggestion: A localized suggestion for recovering from the error
- helpAnchor: A localized string that can be used to get help

Let's look at how we can use the LocalizedError protocol by expanding on our PlayerNumberError enumeration. The following example adds the errorDescription property in order to provide a better description of the error:

```
enum PlayerNumberError: Error, LocalizedError {
    case numberTooHigh(description: String)
    case numberTooLow(description: String)
    case numberAlreadyAssigned
    case numberDoesNotExist

    var errorDescription: String? {
        switch self {
        case .numberTooHigh(description: let description):
            return description
        case .numberTooLow(description: let description):
            return description
        case .numberAlreadyAssigned:
            return "Player number already assigned"
        case .numberDoesNotExist:
            return "Player number does not exist"
        }
    }
}
```

Here, notice that we added the `LocalizedError` protocol to the list of protocols that the enumeration conforms to. We then add the `errorDescription` property to the enumeration, which adds a description to each of the error cases.

We can now use this as shown in the following code:

```
do {
    let player = try myTeam.getPlayerByNumber(number: 34)
    print("Player is \(player.firstName) \(player.lastName)")
} catch let error as PlayerNumberError {
    print("Error: " + error.localizedDescription)
}
```

In this code, we now catch the error and typecast it using the as keyword, and then use the `localizedDescription` property when we print out the error message. The output of this code is:

```
Error: Player number does not exist
```

Now let's look at how we can use the `defer` function to perform a cleanup if an error occurs.

Defer functions

If we need to perform a cleanup action, regardless of whether we have any errors, we can use a defer statement. We use `defer` statements to execute a block of code just before the code execution leaves the current scope. The following example shows how we can use the `defer` statement:

```
func deferFunction() throws {
    print("Function started")
    var myTeam = BaseballTeam()
    var str: String?
    defer {
        print("In defer block")
        if let s = str {
            print("str is \(s)")
        }
    }
    str = "Test String"
    try myTeam.addPlayer(player:("David", "Ortiz", 34))
    print("Function finished")
}
```

If we were to call this function, the first line to be printed to the console would be `Function started`. The code's execution would skip over the `defer` block and execute the next two lines of code. If no error was thrown from the `addPlayer()` function, then `Function finished` would be outputted to the console. Whether the `addPlayer()` function threw an error or not, the `defer` block of code would be executed just before we leave the function's scope, and we would see the `In defer block` message in the console. The following is the output of this function if there were no errors thrown from the `addPlayer()` function:

```
Function started
Function finished
In defer block
str is Test String
```

If an error was thrown from the `addPlayer()` function, the output would look like this:

```
Function started
In defer block
str is Test String
```

Notice in the output where the error was thrown that `Function finished` was never outputted to the screen; however, the `defer` block was still called.

The `defer` block will always be called before the execution leaves the current scope, even if an error is thrown, as shown in this example. The `defer` statement is very useful when we want to make sure we perform all the necessary cleanup, even if an error is thrown. For example, if we successfully open a file to write to, we will always want to make sure we close that file, even if we encounter an error during the write operation. In this case, we could put the file-closed functionality in a `defer` block, ensuring that the file is always closed prior to leaving the current scope.

Multi-pattern catch clauses

Earlier in the chapter, we had code that looked like this:

```
do {
    try myTeam.addPlayer(player:("David", "Ortiz", 34))
} catch PlayerNumberError.numberTooHigh(let description) {
    print("Error: \(description)")
} catch PlayerNumberError.numberTooLow(let description) {
    print("Error: \(description)")
```

```
    } catch PlayerNumberError.numberAlreadyAssigned {
        print("Error: number already assigned")
    } catch {
        print("Error: Unknown Error")
    }
```

In this code, you may have noticed that the catch clause for the PlayerNumberError.numberTooHigh and PlayerNumberError.numberTooLow errors contains duplicate code. As a developer who may have to maintain this code, it is always good to find ways to eliminate duplicate code like this. In Swift, we can eliminate duplicate code like this by utilizing the multi-pattern catch clauses. Let's look at this by rewriting the previous code to use a multi-pattern catch clause:

```
    do {
        try myTeam.addPlayer(player:("David", "Ortiz", 34))
    } catch PlayerNumberError.numberTooHigh(let description),
    PlayerNumberError.numberTooLow(let description) {
        print("Error: \(description)")
    } catch PlayerNumberError.numberAlreadyAssigned {
        print("Error: number already assigned")
    } catch {
        print("Error: Unknown Error")
    }
```

Notice that in the first catch clause, we are now catching both the PlayerNumberError. numberTooHigh and PlayerNumberError.numberTooLow error. The patterns to match are separated by a comma in the catch clause.

Now, let's take a look at typed throws.

Typed throws

In previous versions of Swift, functions could only declare that they would throw errors without specifying the error type(s). This led to less precise error handling, especially with frameworks, because the receiving function did not know the error types that could be thrown. Now, with typed throws, you can define functions that throw specific error types.

Let's look at how typed throws work by looking at the addPlayer() function that we defined in our BaseballTeam type earlier in the chapter. The addPlayer() function was previously defined like this;

```
mutating func addPlayer(player: BaseballPlayer) throws { }
```

With this function definition, we were unaware of what types of errors could be thrown. With typed throws, we can define this function like this:

```
mutating func addPlayer(player: BaseballPlayer) throws(PlayerNumberError)
{ }
```

Typed throws are not required, and untyped throws can still be defined. The following shows the three ways functions are defined internally with typed and untyped errors:

- Functions that have a typed throw will throw an error of that type if an error is thrown.
- A function that has an untyped throw is the same as a typed throw, but internally, that error type will be defined as the any type.
- A function that does not throw an error has an error type defined as Never.

Now that we have seen how to use error handling with Swift, let's look at how to use the availability attribute with Swift.

The availability attribute

Developing our applications for the latest **Operating System** (**OS**) version gives us access to all the latest features for the platform that we are developing for. However, there are times when we want to also target older platforms. Swift allows us to use the availability attribute to safely wrap code to run only when the correct version of the operating system is available.

> The availability attribute is only available when we use Swift to develop for Apple platforms, and is not available for Linux or Windows development.

The availability attribute essentially lets us run a block of code, if we are running on the specified version of the operating system or higher; otherwise, run another block of code.

There are two ways in which we can use the availability attribute. We can use an `if` or a `guard` statement with the availability attribute to run a specific block of code, or we could use the availability attribute to mark a method or type as available only on certain platforms.

The availability attribute accepts up to six comma-separated arguments, which allows us to define the minimum version of the operating system or application extension needed to execute the code. These arguments are as follows:

- `iOS`: This is the minimum iOS version that is compatible with our code.
- `OSX`: This is the minimum OS X version that is compatible with our code.
- `watchOS`: This is the minimum watchOS version that is compatible with our code.
- `tvOS`: This is the minimum tvOS version that is compatible with our code.
- `iOSApplicationExtension`: This is the minimum iOS application extension that is compatible with our code.
- `OSXApplicationExtension`: This is the minimum OS X application extension that is compatible with our code.

After the argument, we specify the minimum version that is required. We only need to include the arguments that are compatible with our code. As an example, if we are writing an iOS application, we only need to include the iOS argument in the availability attribute. We end the argument list with an * (asterisk) as it is a placeholder for future versions.

Let's look at how we would execute a specific block of code only if we met the minimum requirements:

```
if #available(iOS 16.0, OSX 14.10, watchOS 9, *) {
    //Available for iOS 16, OSX 14.10, watchOS 9 or above
    print("Minimum requirements met")
} else {
    //Block on anything below the above minimum requirements
    print("Minimum requirements not met")
}
```

In this example, the `if #available(iOS 16.0, OSX 14.10, watchOS 9, *)` line of code prevents the block of code from executing when the application is run on a system that does not meet the specified minimum operating system version. In this example, we also use the `else` statement to execute a separate block of code if the operating system does not meet the minimum requirements.

We can also restrict access to a function or a type. In the previous code, the `available` attribute was prefixed with the # character. To restrict access to a function or type, we prefix the `available` attribute with an @ character. The following example shows how we could restrict access to a type and function:

```
@available(iOS 16.0, *)
    func testAvailability() {
        // Function only available for iOS 16 or above
}

@available(iOS 16.0, *)
    struct TestStruct {
        // Type only available for iOS 16 or above
}
```

In the previous example, we specified that the `testAvailability()` function and the `testStruct()` type can only be accessed if the code was run on a device that has iOS version 16 or newer. In order to use the `@available` attribute to block access to a function or type, we must wrap the code that calls that function or type with the `#available` attribute.

The following example shows how we could call the `testAvailability()` function:

```
if #available(iOS 16.0, *) {
    testAvailability()
} else {
    // Fallback on earlier versions
}
```

In this example, the `testAvailability()` function is only called if the application is running on a device that has iOS version 16 or later.

Now, let's look at the inverse of the availability attribute: the unavailability attribute.

Unavailability

Where the availability attribute lets us run a block of code if we are running on the specified version of the operating system or higher, the unavailability attribute lets us run a block of code if we are not running on the specified version of the operating system or higher.

Prior to the unavailability attribute, we may have had code similar to this:

```swift
if #available(iOS 16, *) { } else {
    // Functionality to make iOS 15 and earlier work
}
```

This would be used when we needed code to make the functionality of our application work when it is being run on an older version of the operating system. Now, with the unavailability attribute, we can write the code like this:

```swift
if #unavailable(iOS 16) {
    // Functionality to make iOS 15 and earlier work
}
```

An important difference between the availability and unavailability attributes is that the platform wildcard * is not allowed with the unavailability attribute. This is to avoid issues about which platforms should be considered unavailable. The unavailability attribute will only check specific platforms within the list. Here's an example of using the unavailability attribute with multiple platforms:

```swift
if #unavailable(iOS 16, watchOS 9) {
    // Functionality for iOS 15, watchOS 8, and earlier
}
```

The unavailability attribute makes your code more readable and easier to maintain, especially when dealing with platform-specific logic.

Summary

In this chapter, we looked at Swift's error-handling mechanisms. In Swift, we define errors using types that conform to the Error protocol, often represented by enumerations for clarity and conciseness. These error types encapsulate specific error conditions, aiding in code maintenance and comprehension.

We showed how to throw errors from functions using the throws keyword and catch them using the do-catch block. Error patterns are matched within catch clauses, allowing for precise error handling. Additionally, Swift provides a defer statement for executing cleanup code regardless of whether an error occurred.

This chapter also explored Swift's availability and unavailability attributes, enabling conditional execution of code based on the platform's version. Additionally, we looked at the `LocalizedError` protocol and typed throws.

In the next chapter, we will look at Swift's built-in regular expression functionality.

9

Regular Expressions

Regular expressions, commonly referred to as regex or regexp, are powerful tools used in software development to search for and manipulate text based on patterns. A regular expression is essentially a sequence of characters that defines a specific search pattern. This pattern can then be used to efficiently search, match, and manipulate strings of text within Swift code.

These patterns, constructed with literal characters, character classes, and quantifiers, among other elements, enable developers to express complex matching criteria concisely. From validating user inputs, such as email addresses and phone numbers, to parsing structured data formats, such as CSV files or log entries, regular expressions offer a very powerful solution. Regular expressions may seem a little confusing at first; however, once we start to use them, we begin to appreciate their efficiency and flexibility. With regular expressions, tasks that would otherwise require extensive manual string manipulation or custom parsing algorithms can often be accomplished with just a few lines of code.

We will cover the following topics in this chapter:

- The building blocks of regular expressions
- How to use regular expression literals
- Using the Regex type in Swift
- Using Regex builders

At its core, regular expression parsing revolves around interpreting and applying patterns to identify, search through, and manipulate text. The process of parsing involves analyzing a string against a given regular expression pattern to determine whether and where it matches. This enables a wide range of operations, from simple tasks, such as finding all occurrences of a word, to more complex tasks, such as validating the structure of a phone number.

Understanding regular expression parsing requires familiarity with the syntax and semantics of regular expressions themselves. This includes knowledge of metacharacters, quantifiers, character classes, and other constructs that define the rules for pattern matching. Let's look at the building blocks of a regular expression.

How regular expressions are built

Regular expressions are constructed from a combination of elements, each with a specific purpose.

Literals

These are characters or sequences of characters that are matched exactly as they appear in the regular expression. For example, the regex abc will match the sequence "abc" within a string.

Metacharacters

These are special characters that have a reserved meaning within regular expressions. Some common metacharacters are:

- . (dot): This matches any single character except newline.
- \: This escapes a metacharacter, allowing it to be treated as a literal.
- [ ]: This matches any one of the characters inside the brackets.

Quantifiers

These specify how many occurrences of the preceding element are allowed. They include:

- *: This matches zero or more occurrences.
- +: This matches one or more occurrences.
- ?: This matches zero or one occurrence.
- {n}: This matches exactly n occurrences.
- {n, }: This matches n or more occurrences.
- {n,m}: This matches between n and m occurrences.

Anchors

These are used to specify the position within the string where a match can begin or end. Common anchors include:

- ^: This matches the start of the string.
- $: This matches the end of the string.
- \b: This matches a word boundary.

Modifiers

These affect how the pattern is matched. Common modifiers include:

- i: Case-insensitive matching
- g: Global matching, which finds all matches rather than stopping after the first match
- m: Multiline matching, which treats the beginning and end characters (^ and $) as working with each line rather than the whole string

Character classes

These enclose a group of characters that you want to match. For example, "[a-z]" will match any lowercase letter from "a" to "z". As shortcuts, we may use:

- \d: This matches any digit.
- \w: This matches any word character.
- \s: This matches any whitespace character.

Grouping and capturing

Parentheses, ( ), are used to group parts of an expression together, treating it as a single unit. They may also capture the matched text for later reference.

Assertions

Assertions in Swift's regular expression engine enable us to verify that a certain condition exists at a specific position in a string, without consuming any characters. This makes them very useful when we want to match text based on what appears before or after a certain point, while keeping that surrounding context out of the match itself.

There are two main categories of assertions: lookaheads and lookbehinds. A lookahead assertion checks for a condition that must or must not appear immediately after the current position in the string. Swift supports both positive lookaheads, written as (?=...), which ensure a specific pattern follows, and negative lookaheads, written as (?!...), which ensure a specific pattern does *not* follow.

A lookbehind assertion evaluates what comes before the current position. Positive lookbehinds, written as (?<=...), assert that a certain pattern must precede the match, while negative lookbehinds, written as (?<!...), assert that a certain pattern must *not* precede it. Like lookaheads, these perform their checks without consuming any characters in the process.

Assertions are especially valuable when we need to ensure a value appears after a specific prefix, or confirming that a certain word is not followed or preceded by another.

Now that we have a list of the elements that make up a regular expression, let's start exploring Swift's support for regular expressions with regular expression literals.

Regular expression literals

One of the easiest ways to use regular expressions is with regular expression literals. With regular expression literals, we can create instances directly in our code using the / delimiter:

```
let pattern = /\b\w+\b/
text = "Hello from regex literal"
let matches = text.matches(of: pattern)
for match in matches {
    print("-- \(text[match.range])")
}
```

Here, we begin by creating a regular expression literal. The pattern /\b\w+\b/ will match each word in a string. We use the matches(of:) method to return an array of matches within the string, and then the for-in loop is used to loop through each match and output it to the console.

> What makes regular expression literals even more powerful is that we can use them in string methods such as range(of:), replacing(_, with:), wholeMatch(of:), and trimmingPrefix().

Instead of the matches() method that we've just looked at, we can also use the Regex type, which is part of the Swift standard library.

Regex type

With the Regex type in the Swift standard library, we can search for a pattern in a string and then use methods such as contains(:), firstMatch(of:), or matches(of:) in order to find matches. We can also use instances of the Regex type with the string matches(of:) method that we saw in the previous section. Let's look at how we can use the Regex type, starting with the string methods.

To see how we can use the Regex type with our string methods, we will use the same example that we used in the previous section; however, rather than creating a regular expression literal, we will create an instance of the Regex type. Here is the code:

```
let pattern = Regex(/\b\w+\b/)
text = "Hello from regex literal"
let matches = text.matches(of: pattern)
for match in matches {
    print("-- \(text[match.range])")
}
```

This code begins by creating an instance of the `Regex` type using the pattern `/\b\w+\b/`, which will match each word in a string. We use the `matches(of:)` method of the string to return an array of matches within the string, and then the `for-in` loop is used to loop through each match and output it to the console.

In that example, there wasn't really any difference between the regular expression literal and the `Regex` type, and in a lot of ways, they can be used interchangeably.

> In examples like this, the choice of method is really up to the developer's discretion. Personally, I prefer using the `Regex` type because I find it easier to quickly understand what the code is doing, while others might prefer the brevity of using the `matches()` method.

Let's look at another example. In this example, we will look at how to validate an email address, starting with the `Regex` example:

```
let pattern = Regex(
    /\b[A-Z0-9._%+-]+@[A-Z0-9.-]+\.[A-Z]{2,}\b/).ignoresCase()
let address = "hoffman.jon@mydomain.com"
let match = address.wholeMatch(of: pattern)
```

In this example, we are using the `[A-Z0-9._%+-]+@[A-Z0-9.-]+\.[A-Z]{2,}` regular expression to validate the email address. We are also using the `wholeMatch(of:)` method to see whether the whole string matches the regular expression. We could rewrite this code to use regular expression literals like this:

```
let pattern = /\b[A-Z0-9._%+-]+@[A-Z0-9.-]+\.[A-Z]{2,}\b/.ignoresCase()
let address = "hoffman.jon@mydomain.com"
let match = address.wholeMatch(of: pattern)
```

Both code examples would work as expected and the `match` constant would contain an optional if the `address` string contained a valid email address.

When we import the `RegexBuilder` module, we can use it to create `Regex` instances using the clear and flexible declarative syntax of `RegexBuilder`. Let's take a look at `RegexBuilder`.

RegexBuilder

Regular expressions, while extremely powerful, are often cryptic and difficult to read, as we can see from the examples we've looked at so far in this chapter. They can be especially hard for those new to regular expressions or not very familiar with their syntax. `RegexBuilder` addresses this by allowing us to construct regular expressions using a domain-specific language, making them easier to read and more maintainable.

`RegexBuilder` has several elements, defined in the `RegexBuilder` module. Some of the more popular ones are:

- **Capture**: This captures a group of characters.
- **ZeroOrMore**: This matches zero or more occurrences of the preceding element.
- **OneOrMore**: This matches one or more occurrences of the preceding element.
- **Optional**: This matches the preceding element zero or one time.
- **Sequence**: This matches a sequence of elements.
- **Alternation**: This matches one of the provided alternatives.
- **.word**: This matches a word character.
- **.whitespace**: This matches a whitespace character.

> These elements can match as many occurrences as possible or as few as possible, depending on how they are used.

Let's see how `RegexBuilder` works by converting the regular expression examples we looked at earlier in this chapter, which matched each word, with `RegexBuilder`:

```
let pattern = Regex {
    Anchor.wordBoundary
    OneOrMore(.word)
    Anchor.wordBoundary
}

let str = "Hello from RegexBuilder and Swift"
let matches = str.matches(of: pattern)
```

```
for match in matches {
    print("++ \(str[match.range])")
}
```

If we ran this code, the output would look like this:

```
++ Hello
++ from
++ RegexBuilder
++ and
++ Swift
```

If we compare the `Regex` instance in this example to the regular expression that we used in previous examples, \b\w+\b, we can see there is a one-to-one match with the elements:

```
\b  = Anchor.wordBoundary
\w+ = OneOrMore(.word)
\b  = Anchor.wordBoundary
```

We do not have to match these one to one; another way to get similar results is to write the `RegexBuilder` like this:

```
let namePattern = Regex {
    Capture {
        OneOrMore(.word)
        ZeroOrMore(.whitespace)
    }
}
```

In this example, we are using the `Capture` element to capture one or more words and zero or more whitespaces after the word. The output from this example would produce very similar output to the previous example, except that we would capture the spaces as well.

Let's look at a more complex example, one that would validate email addresses. We would write a `RegexBuilder`, as shown in the next example:

```
let pattern = Regex( Regex {
    Anchor.wordBoundary
    OneOrMore {
        CharacterClass(
            .anyOf("._%+-"),
            ("A"..."Z"),
```

```
                ("0"..."9")
            )
        }
        "@"
        OneOrMore {
            CharacterClass(
                .anyOf(".-"),
                ("A"..."Z"),
                ("0"..."9")
            )
        }
        "."
        Repeat(2...) {
            ("A"..."Z")
        }
        Anchor.wordBoundary
}).ignoresCase()
```

In this example, we use the `Anchor.wordBoundary` elements to wrap our validation code. We then look to match each section of the email address just like we did with the regular expression string: \b[A-Z0-9._%+-]+@[A-Z0-9.-]+\.[A-Z]{2,}\b. In the regular expression string, each section was separated with the + (plus) symbol.

> You can use the foundation date, number, currency, and URL parses with `RegexBuilder`. Don't try to write your own regex to match dates.

Now, let's see how we can convert regular expressions to `RegexBuilder` format.

Converting regular expressions to RegexBuilder format

As we can see, the `RegexBuilder` code is much easier to read and understand, which makes it also easier to maintain and troubleshoot. If you are familiar with regular expressions but would like to use `RegexBuilder`, Xcode has a handy conversion tool that will automatically convert your regular expressions to `RegexBuilder` code.

To use Xcode's built-in conversion tool, simply right-click on the regular expression you would like to use, select **Refactor**, and then select **Convert to Regex Builder**. The following screenshot illustrates this.

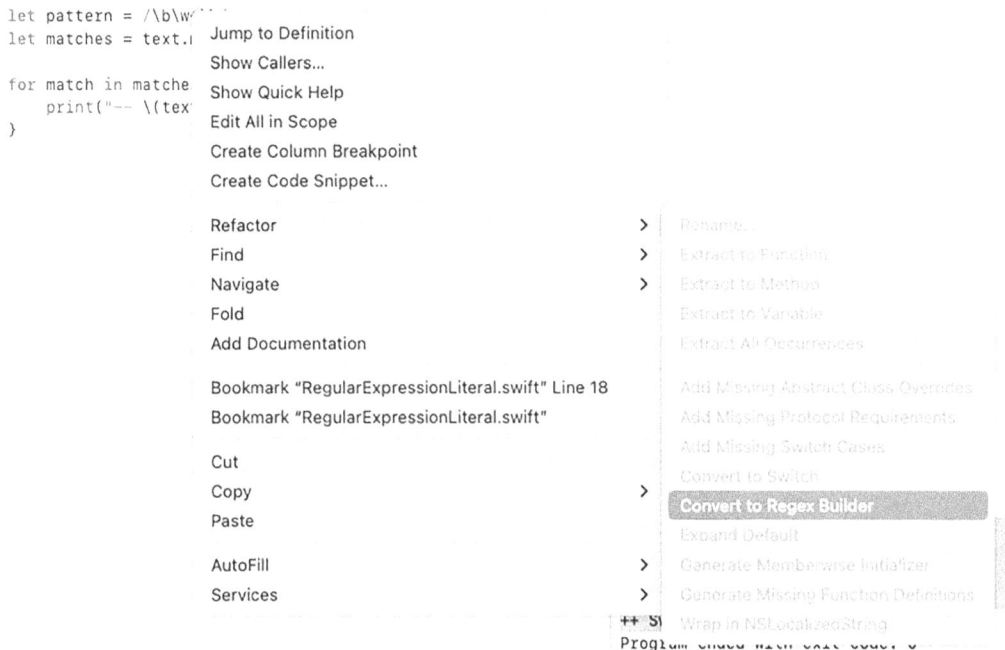

Figure 9.1: Converting a regular expression to Regex Builder format in Xcode

Once you select **Convert to Regex Builder**, Xcode will convert the regular expression to `RegexBuilder` code.

Now, let's look at some additional functionality that we get when we use `RegexBuilder` as compared to regular expressions, starting with the ability to transform its matches.

Transform a match with RegexBuilder

With `RegexBuilder`, we are able to apply transformations to the matches that it finds; for example, if `TryCapture` is used rather than `Capture`, then the whole match will fail if the capture fails. Let's see how this works with the following code:

```
let str = "I am a dog who is 1 years old and my name is Luna"
let pattern = Regex {
    "I am a "

    Capture {
```

```
        OneOrMore(.word)
    }

    " who is "

    TryCapture {
        OneOrMore(.digit)
    } transform: { match in
        Int(match)
    }

    " years old and my name is "

    Capture {
        OneOrMore(.word)
    }
}

let matches = str.matches(of: pattern)
for match in matches {
    print("-- \(str[match.range])")
}
```

In the `RegexBuilder` code, we are attempting to match a specific string with three string literals, " I am a "," who is ", and " years old and my name is ", and three capture blocks. The middle capture block uses a `TryCapture` statement and attempts to convert the match to an integer. If the code does not find a numeric digit there, the match will fail. If we ran the code as shown, we would find a match; however, if we change the `str` constant to I am a dog who is one year old and my name is Luna (replacing the "1" with "one"), the match would fail.

While this code in itself may not seem that exciting, where it really comes into play is when we wish to use name references to capture the matches into variables.

Capture a match with RegexBuilder

`RegexBuilder` allows us to use the `Reference` type to capture values and access them later. This can be very useful for complex patterns where we would like to access specific portions of the match.

Let's see how this works using the same example as we used in the previous section:

```swift
let animalTypeRef = Reference(Substring.self)
let ageRef = Reference(Int.self)
let nameRef = Reference(Substring.self)

let str = "I am a dog who is 1 year old and my name is Luna"
let pattern = Regex {
    "I am a "

    Capture(as: animalTypeRef) {
        OneOrMore(.word)
    }

    " who is "

    TryCapture(as: ageRef) {
        OneOrMore(.digit)
    } transform: { match in
        Int(match)
    }

    " years old and my name is "

    Capture(as: nameRef) {
        OneOrMore(.word)
    }
}

let matches = str.matches(of: pattern)
for match in matches {
    print("- Animal Type:  \(match[animalTypeRef])")
    print("- Name: \(match[nameRef])")
    print("- Age: \(match[ageRef])")
}
```

In this code, we start by defining three references named `animalTypeRef`, `ageRef`, and `nameRef`. We define `ageRef` as an integer type and the other two references as substring types. In the code, within the `Capture` and `TryCapture` calls, we use the as keyword to capture the elements within the references. Finally, as we loop through the matches, we are able to use the references to extract the data. As we can see, this makes it incredibly easy to extract specific data from the matches.

Summary

Regular expressions are an indispensable tool for pattern matching and manipulation in text. With their concise syntax and powerful capabilities, they enable a wide range of tasks, from simple string searches to complex data extraction. In this chapter, we explored the core concepts of regular expressions, including their building blocks.

We examined Swift's support for regular expressions, starting with the use of regular expression literals and the `Regex` type. Both methods allow for pattern matching and manipulation directly within Swift code. Next, we introduced `RegexBuilder`, which provides a more readable and maintainable way to define regular expressions using a declarative syntax.

Finally, at the end of the chapter, we looked at the advanced features of `RegexBuilder`, including transforming and capturing matches using references.

In the next chapter, we'll look at how we can create subscripts for our custom types.

Unlock this book's exclusive benefits now

Scan this QR code or go to `packtpub.com/unlock`, then search this book by name.

Note: Keep your purchase invoice ready before you start.

10

Custom Subscripting

Swift provides various ways to customize and extend the behavior and functionality of custom types. One very powerful feature is custom subscripting, which enables developers to define unique subscripting behavior for their custom types. This capability makes accessing elements within an instance of these types more intuitive and expressive.

We will examine the concepts of custom subscripting in Swift in this chapter. We will explore how custom subscripting can simplify API design, making it easier to interact with custom types. Whether you're working with collections, custom data models, or other complex data structures, mastering custom subscripting can significantly enhance the clarity and usability of your code.

In this chapter, we will cover the following topics:

- What are custom subscripts?
- Adding custom subscripts to classes and structures
- Creating read/write and read-only subscripts
- Using external names without custom subscripts
- Using multidimensional subscripts

Introducing subscripts

In Swift, subscripts provide a convenient way to access elements of a collection, list, or sequence. They enable us to set or retrieve values by index or key, bypassing the need for traditional getter and setter methods. When used correctly, subscripts can greatly enhance the usability and readability of our custom types.

Swift enables us to define multiple subscripts for a single type. When we define multiple subscripts, the appropriate subscript is selected based on the parameter type or external names. Additionally, we can set external parameter names for our subscripts to differentiate between subscripts that use the same index type.

Custom subscripts function similarly to those in arrays and dictionaries. For instance, accessing an element in an array uses the array[index] syntax. When we define a custom subscript for our types, we can access elements using the same ourType[key] syntax.

To ensure custom subscripts are effective, they should integrate seamlessly with the class, structure, or enumeration they belong to. While subscripts can enhance code usability and readability, overuse can make them feel unnatural and difficult to understand.

In this chapter, we will explore several examples of creating and using custom subscripts. We'll start by reviewing how subscripts work with Swift arrays, helping us to understand their use within the language. By following Apple's implementation patterns, we can make our custom subscripts intuitive and easy to use.

Subscripts with Swift arrays

The following example shows how to use subscripts to access and change the values of an array:

```
var arrayOne = [1, 2, 3, 4, 5, 6]
print(arrayOne[3])  //Displays '4'
arrayOne[3] = 10
print(arrayOne[3])  //Displays '10'
```

In the preceding example, we create an array of integers and then use the subscript syntax to display and change the element at index three.

Subscripts are mainly used to set or retrieve information from a collection. We generally do not use subscripts when specific logic needs to be applied to determine which item to select. As an example, we would not want to use subscripts to append an item to the end of the array or to retrieve the number of items in the array.

> To append an item to the end of an array, or to get the number of items in an array, we use functions or properties, as shown in the following code:
> ```
> arrayOne.append(7)  //append 7 to the end of the array
> arrayOne.count  //returns the number of items in an array
> ```

Subscripts in our custom types should follow the same standard set by the Swift language so that other developers who use our types are not confused by the implementation. The key to knowing when to use subscripts, and when not to, is to understand how they will be outside of the custom type.

Creating and using custom subscripts

Let's look at how to define a subscript that is used to read and write to a backend array. Reading and writing to a backend storage class is one of the most common uses of custom subscripts. However, as we will see in this chapter, we do not need to have a backend storage class because subscripts conceal how the data is stored. The data can be stored in a backend storage class, array, dictionary, or even computed property, and subscripts provide a universal way to access it.

The following code shows how to use a subscript to read and write to an array:

```swift
class MyNames {
    private var names = ["Jon", "Kailey", "Kai"]
    subscript(index: Int) -> String {
        get {
            names[index]
        }
        set {
            names[index] = newValue
        }
    }
}
```

As we can see, the syntax for subscripts is similar to how we define properties within a class using the get and set keywords. The difference is that we declare the subscript using the subscript keyword. We then specify one or more inputs and the return type.

We can now use the custom subscript just as we would use subscripts with arrays and dictionaries. The following code shows how to use the subscript in the preceding example:

```swift
let name = MyNames()
print(name[0])  //Displays 'Jon'
name[0] = "Kailey"
print(name[0])  //Displays 'Kailey'
```

In the preceding code, we create an instance of the MyNames class and display the original name at index 0. We then change the name at index 0 and redisplay it. In this example, we use the subscript that is defined in the MyNames class to retrieve and set elements of the names array within the class.

While it is possible to expose the names array and make it available for external code to access directly, this would lock our code into using an array to store the data. In the future, if we wanted to switch the backend storage mechanism to a dictionary object, or even an SQLite database, we would have difficulty doing so because the external code that used this type would be reliant on the array. Subscripts are excellent at abstracting the underlying storage details of our custom types, ensuring that external code does not depend on specific storage implementations.

Direct access to the names array would also prevent us from validating the data being inserted by external code. With subscripts, we can add validation to our setters to ensure that the data is correct before adding it to the array. For instance, we could validate that the names contain only valid alphabetic characters and certain special characters that are valid in names. This capability is particularly useful when creating a framework or library, as it helps maintain data integrity.

Read-only custom subscripts

We can make the subscript read-only by not declaring a setter method within the subscript. The following code examples show the two ways to declare a read-only property by not declaring a setter method.

This first example shows how to declare the read-only subscript but not explicitly declare the getter or setter:

```
//No getter/setters implicitly declared
subscript(index: Int) -> String {
    names[index]
}
```

The following example shows how to declare a read-only property by only declaring a getter method:

```
//Declaring only a getter
subscript(index: Int) -> String {
    get {
        names[index]
    }
}
```

In the first example, we do not define either a getter or setter method; therefore, Swift sets the subscript as read-only, and the code acts as if it were in a getter definition. In the second example, we specifically set the code in a getter definition. Both examples are valid read-only subscripts.

Note that write-only subscripts are not valid in Swift.

Calculated subscripts

While the preceding example is very similar to using stored properties, subscripts can be used in a similar manner to computed properties. Let's see how this is done:

```
struct MathTable {
    var num: Int
    subscript(index: Int) -> Int {
        return num * index
    }
}
```

In the earlier example, an array is used as a backend storage mechanism for the subscript. In this example, the return value is calculated using the value of the subscript itself. We would use this subscript as follows:

```
var mathTable = MathTable(num: 5)
print(table[4])
```

This example displays the calculated value of 5 (the number defined in the initialization) multiplied by 4 (the subscript value), which would be 20.

Subscript values

In the preceding subscript examples, the subscripts all accepted integers as the value; however, we are not limited to integers. In the following example, we will use a String type as the value for the subscript. The subscript will also return a value of the String type.

```
struct Hello {
    subscript (name: String) -> String {
        return "Hello \(name)"
    }
}
```

In this example, the subscript takes a string as the value within the subscript and returns a message saying `Hello`. Let's look at how to use this subscript:

```
var hello = Hello()
let greeting = hello["Jon"]
```

In this code, the `greeting` constant would contain the string `Hello Jon`.

Static subscripts

Static subscripts enable us to use the subscript without having to create an instance of the type. This is similar to how we would use static properties or methods. Let's see how this works:

```
struct Hello {
    static subscript (name: String) -> String {
        return "Hello \(name)"
    }
}
```

In the previous code, we create a structure named `Hello` and, within this structure, we define a static subscript using the `static` keyword within the `subscript` declaration. We are now able to use this subscript as follows:

```
let greeting = Hello["Jon"]
```

In the previous code, the `greeting` constant would contain the string `Hello Jon`. Note that we did not have to create an instance of the `Hello` structure to use the subscript.

External names for subscripts

Earlier in the chapter, we mentioned that we could have multiple subscript signatures for our custom types. The appropriate subscript is chosen based on the type of index passed into the subscript. However, there are times when we may wish to define multiple subscripts that have the same type. For this, we could use external names in a similar way to how we define external names for the parameters of a function.

Let's rewrite the original `MathTable` structure from the *Calculated subscripts* subsection to include two subscripts that each accept an integer as the type. However, one will perform a multiplication operation, and the other will perform an addition operation:

```
struct MathTable {
    var num: Int
    subscript(multiply index: Int) -> Int {
        return num * index
    }
    subscript(add index: Int) -> Int {
        return num + index
    }
}
```

As we can see, in this example, we define two subscripts, and each subscript accepts an integer type. The difference between the two subscripts is the external name within the definition. In the first subscript, we define an external name, `multiply`, because we multiply the value of the subscript by the `num` property defined when the `MathTable` structure is initiated. In the second subscript, we define an external name of `add` because we add the value of the subscript to the `num` property within the subscript.

Let's look at how to use these two subscripts:

```
var mathTable = MathTable(num: 2)
print(table[multiply: 5])  //Displays 10 because 5*2=10
print(table[add: 4])  //Displays 6 because 4+2=6
```

If we run this example, we will see that the correct subscript is used, based on the external name within the subscript.

Using external names within our subscript is very useful if we need multiple subscripts of the same type. In Swift, subscripts are designed to provide a concise and universal way to access data that's similar to how we access data in an array. Adding external names can add unnecessary complexity, however; therefore, I do not recommend using external names unless they are needed to distinguish between multiple subscripts.

Multidimensional subscripts

While the most common subscripts are those that take a single parameter, subscripts are not limited to single parameters. They can take any number of input parameters, and these parameters can be of any type. Subscripts that take multiple parameters are called multidimensional subscripts.

Let's look at how we could use a multidimensional subscript to implement a Tic-Tac-Toe board. A Tic-Tac-Toe board looks similar to the following diagram:

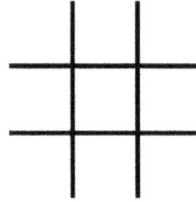

Figure 10.1: Empty Tic-Tac-Toe board

The board can be represented by a two-dimensional array, where each dimension has three elements. The upper-left box of the board can be represented by the coordinates 0,0, while the lower-right box of the board can be represented by the coordinates 2,2. The middle box can have the coordinates 1,1. Each player will take turns placing their pieces (typically an X or an O) onto the board until one player has three pieces in a line or the board is full.

Let's look at how we could implement a Tic-Tac-Toe board using a multidimensional array and multidimensional subscripts:

```
struct TicTacToe {
    var board = [["","",""],["","",""],["","",""]]
    subscript(x: Int, y: Int) -> String {
        get {
            return board[x][y]
        }
        set {
            board[x][y] = newValue
        }
    }
}
```

We start the TicTacToe structure by defining a 3x3 array, also known as a matrix, which will represent the game board. We then define a subscript that can be used to set and retrieve player pieces on the board. The subscript will accept two integer values.

We define multiple parameters for our subscripts by putting the parameters between parentheses. In this example, we are defining the subscript with the parameters (x: Int, y: Int). We can then use the x and y variable names within our subscripts to access the values that are passed in.

Let's look at how to use this subscript to set the user's pieces on the board:

```
var board = TicTacToe()
board[1,1] = "x"
board[0,0] = "o"
```

If we run this code, we will see that we added the x piece to the center square and the o piece to the upper-left square, so our game board will look like the following:

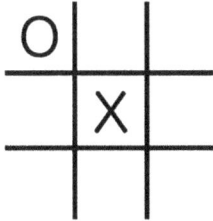

Figure 10.2: Tic-Tac-Toe board with two player pieces

We are not limited to using only one type within our multidimensional subscripts. For example, we could have a subscript of (x: Int, y: Double, z: String).

We can also add external names for our multidimensional subscript types to help identify what the values are used for and to distinguish between subscripts that have the same types. Let's take a look at how to use multiple types and external names with subscripts by creating a subscript that will return an array of string instances based on the values of the subscript:

```
struct SayHello {
    subscript(messageText message: String, messageName name: String,
            number: Int) -> [String]{

        var retArray: [String] = []
        for _ in 0..<number {
            retArray.append("\(message) \(name)")
        }
        return retArray


    }
}
```

In the `SayHello` structure, we define our subscript as follows:

```
subscript(messageText message: String, messageName name: String,
        number: Int) -> [String]
```

This defines a subscript with three elements. Each element has an external name (`messageText`, `messageName`, and `number`) and an internal name (`message`, `name`, and `number`). The first two elements are of the String type and the last one is an Int type. We use the first two elements to create a message for the user that will repeat the number of times defined by the last (`number`) element. We will use this subscript as follows:

```
var message = SayHello()
var myMessage = message[messageText:"Bonjour", messageName:"Jon",5]
```

If we run this code, we will see that the `return` variable contains an array of five strings, where each string has a value of `Bonjour Jon`.

Another use case for subscripts in Swift is to extend types. Let's take a look at how we could do this.

Extending types with subscripts

There are cases where subscripts provided by Swift standard library types don't offer the functionality that we need. In these scenarios, rather than creating a new type, we could extend the existing type by adding a subscript that meets our needs. This can be particularly useful when we need to use elements in a way that the original type doesn't support.

A good example of a type we could extend is the `String` type in Swift, which we could extend to provide access to the individual characters within the String. The following code shows how we could do this with a subscript:

```
extension String {
    subscript(index: Int) -> Character? {
        guard index >= 0 && index < self.count else {
            return nil
        }
        return self[self.index(self.startIndex, offsetBy: index)]
    }
}
```

The extension here adds a subscript to the Swift `String` type that will return the character at the index provided with the subscript. If the index is out of bounds, it will return `nil`. We can now use this extension like this:

```
let myString = "Hello World"
var char = myString[1]
```

In this example, the `char` variable will contain the character e.

Now that we have seen how to use subscripts, let's take a quick look at when not to use custom subscripts.

When not to use a custom subscript

As we have seen in this chapter, creating custom subscripts can really enhance our code. However, we should avoid overusing them or using them in a way that is not consistent with standard subscript usage. The way to avoid overusing subscripts is to examine how subscripts are used in Swift's standard libraries.

Let's look at the following example:

```
struct MyNames {
    private var names:[String] = ["Jon", "Kailey", "Kara"]
    var number: Int {
        get {
            return names.count
        }
    }
    subscript(add name: String) -> String {
        names.append(name)
            return name
    }
    subscript(index: Int) -> String {
        get {
            return names[index]
        }
        set {
            names[index] = newValue
        }
    }
}
```

In the preceding example, within the MyNames structure, we define an array of names that are used within our application. As an example, let's say that, within our application, we display this list of names and allow users to add further names to it. Within the MyNames structure, we define the following subscript, which allows us to append a new name to the array:

```
subscript(add name: String) -> String {
    names.append(name)
    return name
}
```

This would be an example of poor use of subscripts because its usage is not consistent with how subscripts are used within the Swift language itself—we are using it to accept a parameter and add that value. This might cause confusion when the class is used. It would be more appropriate to rewrite this subscript as a function, such as the following:

```
func append(name: String) {
    names.append(name)
}
```

When you are using custom subscripts, make sure that you are using them appropriately.

Summary

In this chapter, we saw that adding support for subscripts to our custom types can greatly enhance their readability and usability. We saw that subscripts can be used to add an abstraction layer between our backend storage class and external code. Subscripts can also be used in a similar manner to computed properties, where the subscript is used to calculate a value.

We looked at examples of multidimensional arrays and how to use external names with subscripts, as well as how to extend types with subscripts. As we noted in the chapter, the key with subscripts is to use them appropriately and in a manner that is consistent with subscripts in the Swift language. It's worth noting that we can also use optionals and generics with subscripts.

In the next chapter, we will look at how to use property observers and wrappers to manage state changes and enhance code reusability.

Unlock this book's exclusive benefits now

Scan this QR code or go to packtpub.com/unlock, then search this book by name.

Note: Keep your purchase invoice ready before you start.

11
Property Observers and Wrappers

Efficiently handling state changes and promoting code reusability are key to building efficient and reliable applications. In Swift, two powerful features that help with these tasks are property observers and property wrappers, which provide effective ways to monitor changes in property values and adjust behaviors.

Property observers enable our applications to keep track of property modifications and react accordingly. They provide a simple and effective way to execute custom actions when the value of a property changes. They enable applications to automatically respond to these changes, ensuring that any necessary updates or validations are performed. Property wrappers, on the other hand, bundle the logic of accessing and modifying a property into reusable components. They simplify our code by handling tasks such as validation or transformation in a centralized place. This not only reduces repetitive code but also makes code easier to read and maintain. By using property wrappers, we can create cleaner and reusable code, making it easier to manage complex state changes and behaviors.

In this chapter, we will cover the following topics:

- What are property observers, and how do we use them?
- What are property wrappers, and how do we use them?

Let's begin by exploring the concept of property observers, their syntax, and practical applications.

Introducing property observers

Property observers in Swift are powerful tools that enable us to execute custom code before or after the value of a property changes, making it easier to implement automatic responses to changing state. They can be added to any non-lazy stored property, making them extremely versatile. Property observers can also be applied to inherited stored properties by overriding them in the subclass, providing a way to customize behavior in derived classes.

Swift provides two kinds of property observers:

- **willSet**: This observer will be called just before the value of a stored property is set. It is passed the new property value as a constant parameter, which by default is named newValue, although you can provide your own name.

- **didSet**: This observer is called immediately after the new value of a stored property is set. It provides a parameter containing the old property value, which by default is named oldValue, which you can use within the observer.

To use a property observer, we define the willSet() and/or didSet() observer within the property's definition, as shown in the following code:

```swift
struct MyStruct {
    var myProperty: String {
        willSet(newName) {
            print("Preparing to change the value of myProperty from \(
                myProperty) to \(newName)")
        }
        didSet {
            if oldValue != myProperty {
                print("The value of myProperty changed from \(
                    oldValue) to \(myProperty)")
            }
        }
    }
}
```

In this code, we define both the `willSet()` and `didSet()` observers for the `myProperty` property; however, the majority of the time, unless we are using the property observers to log state changes, we will use one of the observers, but not both.

In the `willSet()` observer, we defined a custom name for the new value named `newName`; however, we do not need to provide a new name and define it like this: `willSet()`. If we do not provide a new name, the value will be accessible by using the `newValue` constant. Let's look at a real-world example of how we would use a property observer.

> ♡ **Quick tip**: Enhance your coding experience with the **AI Code Explainer** and **Quick Copy** features. Open this book in the next-gen Packt Reader. Click the **Copy** button
>
> **(1)** to quickly copy code into your coding environment, or click the **Explain** button
>
> **(2)** to get the AI assistant to explain a block of code to you.

```
function calculate(a, b) {
  return {sum: a + b};
};
```

> 📖 **The next-gen Packt Reader** is included for free with the purchase of this book. Scan the QR code OR go to `packtpub.com/unlock`, then use the search bar to find this book by name. Double-check the edition shown to make sure you get the right one.

Using property observers

There are many common usages for property observers. Here are a few:

- **Validation**: We can use the `willSet` observer to verify that the new values meet certain conditions before they are set.

- **Updating the UI**: In applications with a user-friendly UI, we can use the `didSet` observer to update the UI based on changes within the backend data.

- **Logging changes**: Property observers can be used to log changes to values for debugging purposes.
- **Cascading changes**: We can modify related properties or state when a property changes.

Let's look at an example where we use a property observer to trigger external events based on changes to the property. In this example, let's imagine we are developing a simple inventory management system for a small online store. Our application will need to monitor the stock level of the products and notify a manager or update the system when stock needs to be reordered.

To do this, we could define our Product type like this:

```
class Product {
    var name: String
    var price: Double
    var stockLevel: Int {
        didSet {
            if stockLevel < minimumStockLevel {
                print("Stock Level is below minimum for \(name)")
                reorderNotification()
            }
        }
    }
    var minimumStockLevel: Int

    init(name: String, price: Double, stockLevel: Int,
        minimumStockLevel: Int) {
        self.name = name
        self.price = price
        self.stockLevel = stockLevel
        self.minimumStockLevel = minimumStockLevel
    }

    // Simulate a sale which reduces stock
    func sell(units: Int) {
        print("Selling \(units) unit(s) of \(name).")
        stockLevel -= units
    }
}
```

In our `Product` type, we define a `didSet()` property observer for the `stockLevel` property. In this observer, we check to see whether the current stock level has fallen below the minimum stock level for that product. If it has, a function is called that will send a notification to reorder the product.

We also define a `sell(units: Int)` method that will be called each time the product is sold. In this function, we decrease the stock level for the product by the number of items sold.

The following code shows how we would use this type:

```
let book = Product(name: "Mastering Swift", price: 34.99, stockLevel: 7,
minimumStockLevel: 5)

book.sell(units: 1)
print("Sold one")
book.sell(units: 1)
print("Sold another")
book.sell(units: 1)
print("Sold a third one")
```

We start this code by defining a product with a stock level of 7 and a minimum stock level of 5. We then call the `sell` method three times, each time selling one item. The results of this code would be:

```
Selling 1 unit(s) of Mastering Swift.
Sold one

Selling 1 unit(s) of Mastering Swift.
Sold another

Selling 1 unit(s) of Mastering Swift.
Stock Level is below minimum for Mastering Swift
Alert: Stock for Mastering Swift needs to be reordered.
Sold a third one
```

In the results, we can see that after we sold three items, the message was triggered to reorder the item because the stock level fell below the minimum stock level.

Now, let's look at how we can use the `willSet` property observer.

One of the most popular use cases for the `willSet` property observer is logging because it enables us to log changes before they occur. Additionally, we can use the `newValue` property to access the new value of the property.

For this example, let's assume we have a `logger()` method that will log any messages we send to it. We could then create a `User` type, as shown here:

```
struct User {
    var userName: String {
        willSet {
            logger("User name changing from \(userName) to \(newValue)")
        }
    }
    var password:  String
}
```

In this code, we use the `willSet` observer to log any changes to the `userName` property prior to setting the value. Within the logger method call, we also use the `newValue` property to log the value that is about to be set.

There are many additional usages for property observers; however, there are certain limitations and considerations to take into account before using them. Here are some of the main ones:

- **Initialization**: Property observers are not called when a property is first initialized. They only monitor changes that occur after the initialization of the property.

- **Performance**: Property observers impact performance because every change triggers the method calls.

- **Complexity**: Overusing property observers can make code harder to follow and maintain because they can introduce complex logic that is not always immediately apparent without inspecting the observers themselves.

Now that we understand how to use property observers, let's look at how we can use property wrappers.

Introducing property wrappers

Property wrappers move the responsibility of reading and writing a property into a separate definition, reducing clutter and duplication in our code. This separation allows the logic for storing, validating, and changing properties to be reused.

Property wrappers can encapsulate common behaviors or functionalities typically associated with property management, such as storage options and data validation, into reusable components. Essentially, property wrappers provide a way to inject custom behavior every time a property is accessed or modified, streamlining property management and enhancing code reusability.

```
@propertyWrapper
struct MyPropertyWrapper<T> {
    private var value: T

    var wrappedValue: T {
        get { /* return the value */}
        set { /* modify the value */ }
    }

    init(wrappedValue initialValue: T) {
        self.value = initialValue
    }
}
```

We define a property wrapper using the `@propertyWrapper` attribute. In our example, we are defining a generic struct named `MyPropertyWrapper` that will work with any type, `T`. We use generics here to allow the wrapper to be used with any data type.

We define a private variable named `value` of type `T`. It holds the underlying value stored within the property wrapper. We define it as private to ensure that it is accessed or modified through the `wrappedValue` computed property.

The `wrappedValue` computed property is defined, which will be used by external code to access the `value` property. This computed property includes both get and set blocks. The get block should contain the code to return the stored value, while the `set` block is used when `wrappedValue` is set and usually modifies the value in some way.

Finally, we have the initializer, which sets up the property wrapper with an initial value. It accepts a single parameter of type `T` and sets the private `value` property to the parameter's value. This is necessary to initialize the property wrapper.

We would use this property wrapper like this:

```
struct MyPropertyWrapperExample {
    @MyPropertyWrapper var number: Int

    init(number: Int) {
```

```
            self.number = number
        }
    }
```

Notice that within the MyPropertyWrapper structure, we use the @MyPropertyWrapper attribute for the number property. Now, let's look at an example of how we could use property wrappers in the real world.

Using property wrappers

Let's look at an example of property wrappers. In this example, we will create a property wrapper that will automatically capitalize the string property that uses this wrapper. The Capitalized property wrapper could be written like this:

```
@propertyWrapper
struct Capitalized {
    private var value: String = ""

    var wrappedValue: String {
        get { value }
        set { value = newValue.capitalized }
    }
}
```

When we were describing the property wrappers, we mentioned that the set block is usually used to modify the values in some way; therefore, within our Capitalized property wrapper here, we use the set block to capitalize the value prior to setting it.

We can now use this property wrapper like this:

```
struct Person {
  @Capitalized var name: String

    init(name: String) {
        self.name = name
    }
}
```

Within our `Person` structure, the `Capitalized` property wrapper is applied by using the @ `Capitalized` attribute before the `name` property. This tells the compiler to use the `Capitalized` property wrapper for this property.

Let's look at how we would use the structure:

```
let person = Person(name: "jon hoffman")
print(person.name)
```

If we ran the following code, we would see that the name is now capitalized even though we did not initialize it that way.

That was a pretty basic usage of a property wrapper. Let's look at a more complex usage where we will validate a number based on a range. We will begin by defining the property wrapper like this:

```
@propertyWrapper
struct ValidateRange {
    private var value: Int
    private let range: ClosedRange<Int>

    var wrappedValue: Int {
        get { value }
        set { value = max(range.lowerBound, min(range.upperBound,
                                            newValue)) }
    }

    init(wrappedValue: Int, _ range: ClosedRange<Int>) {
        self.value = max(range.lowerBound, min(range.upperBound,
                                            wrappedValue))
        self.range = range
    }
}
```

This code defines a property wrapper named `ValidateRange` that is designed to validate that an integer's value is within a specified closed range.

Within the property wrapper, we see a property named range, and it is defined as a `ClosedRange`, which means the range must include both endpoints. By defining the property with the `let` keyword, we are defining it as immutable once it is set.

We are using the range property within our `set` block to ensure that the value being set is within the range defined. Within the `set` block, if `newValue` is less than `range.lowerBound`, it sets the value to `range.lowerBound`. If `newValue` is greater than `range.upperBound`, it sets the value to `range.upperBound`. Otherwise, it sets the value to `newValue`.

We could use this property wrapper as shown in the following code:

```
struct Item {
    @ValidateRange(1...100) var quantity = 5
}
```

There are two things to notice with this code. The first is that we are defining the range with the `@ValidateRange` attribute, which enables us to define a different range for each property that uses this wrapper. Second, we are defining an initial value for the `quantity` property. This is because our initializer requires both the range and the value to be set.

If we ran the following code, we would see that the second line that is outputted is `100`, and not `1000` as we would expect. This is because we defined the range for our quantity property to be between 1 and 100:

```
var item = Item(quantity: 10)
print(item.quantity)
item.quantity = 1000
print(item.quantity)
```

Another capability of property wrappers is the ability to use projected values. Let's take a look at this.

Projected values

Projected values are additional values that a property wrapper can expose. This enables us to add additional logic to a value without changing the value itself.

With projected values, we are able to access the main property through the property's normal name and then, if the projected values are implemented, access the projected value using `$` followed by the property's name.

Let's see how this works. Let's say that we have a `Person` type in our code, as shown here:

```
struct Person {
    var name: String
    var birthDate: Date
}
```

With that code, we could create an instance of the `Person` type and print out the `birthDate` property:

```
let baby = Person(name: "Jon", birthDate: Date())
print(baby.birthDate)
```

This would print out the `birthDate` property in ISO 8601 format, which is YYYY-MM-DD HH:MM:SS +0000, like this:

```
2024-11-12 08:20:23 +0000
```

Ensuring that we capture and preserve the date and time of the person's birth is what we want. Maybe we would also like a way to display only the date. For this, we could create a property wrapper as follows:

```
@propertyWrapper
struct DateFormat {
    var wrappedValue: Date
    var projectedValue: String {
        let formatter = DateFormatter()
        formatter.dateFormat = "yyyy-MM-dd"
        return formatter.string(from: wrappedValue)
    }
}
```

Notice that in this property wrapper, we introduced an additional property named `projectedValue`. By using this name, we are able to access the property using $ and the name of the property. We would then use it as we would any other property wrapper.

Let's now add this to the `birthDate` property of our `Person` type:

```
struct Person {
    var name: String
    @DateFormat var birthDate: Date
}
```

Now, we can access both the original value of the `birthDate` property and a formatted version that just shows the date, as shown with the following code:

```
let baby = Person(name: "Jon", birthDate: Date())
print(baby.birthDate)
print(baby.$birthDate)
```

This code would have the following output.

```
2024-11-12 08:20:23 +0000
2024-11-12
```

Property wrappers abstract the responsibility of reading and writing a property into a separate definition, making your code easier to maintain and reducing redundancy. As you become more familiar with property wrappers, you'll find them indispensable for writing easy-to-maintain code.

Observation of values

Swift 6.2 adds a flexible new way to watch for changes in our data, outside of SwiftUI. With SE-0475, Swift introduces an Observations struct, which we create with a closure. Inside the closure, we can access @Observable properties, and in return, we get an AsyncSequence that emits new values anytime those properties change.

To use this, we start by marking a class as observable using the @Observable macro, which enables any properties that can change to be tracked.

```
@Observable
class Unit {
    var hitPoints = 100
}
```

With this, any code that observes any of the properties of this class will get notified when it changes. Now, to react to those changes, we need to wrap the code that reads the property we wish to observe in an Observations block. That gives us an AsyncSequence that we can loop over with for await:

```
let myUnit = Unit()
let hitPointsRemaining = Observations { myUnit.hitPoints }

for i in 1...5 {
    Task {
        try? await Task.sleep(for: .seconds(2))
        myUnit.hitPoints -= 10
    }
}

for await hitPoints in hitPointsRemaining {
    print("new hit points: \(hitPoints)")
}
```

In this example, the unit loses 10 hit points every two seconds. The Observations block picks up on each change and pushes the new value into the async loop, where we can respond to it. Using @Observable makes it a lot easier to write responsive, event-driven code without having to manage Combine publishers or having to manually setup observation logic.

Summary

Swift's property observers and property wrappers are powerful tools that enable our applications to automatically respond to changes in property values. Property observers enable our application to trigger specific actions when property values are set, thereby improving the application's responsiveness and making code easier to maintain. They are particularly useful for tasks such as data validation, UI updates, and debugging.

Property wrappers abstract the responsibility of managing property access into reusable types, reducing complexity and promoting cleaner, more readable code. They can encapsulate common behavior such as data validation, local storage interaction, or UI updates, making them invaluable for writing easy-to-maintain code. With the use of property wrappers, we can focus on the higher-level logic of our applications, while the wrappers handle the details of property management.

These features not only streamline the development process but also open up possibilities for more efficient and effective code. By using these tools, developers ensure their applications are easier to maintain and scalable.

In the next chapter, we will look at key paths and dynamic member lookup.

12

Dynamic Member Lookup and Key Paths

Dynamic member lookup is an important feature in Swift that enables developers to access properties and methods of a type dynamically using dot syntax, even if those properties and methods are not explicitly defined in the type's declaration. By leveraging dynamic member lookup, we can write more flexible and expressive code.

Swift's key path feature is a powerful and flexible way to access properties of a type indirectly, enabling developers to write reusable and easier-to-maintain code. Key paths are type-safe pointers to properties within an object that enable us to reference and manipulate values dynamically without knowing them at compile time. Key paths can be used with functions such as **map**, **filter**, and **reduce** to perform operations on elements of a collection in a concise manner.

While dynamic member lookup and key paths both provide ways to interact with properties of types flexibly and dynamically, they serve different purposes and have distinct implementations.

In this chapter, we will cover the following topics:

- What is dynamic member lookup, and how do we use it?
- What are key paths, and how do we use them?

Let's start off by looking at dynamic member lookup.

Dynamic member lookup

Dynamic member lookup enables a call to a property that will be dynamically resolved at runtime. This may not make a lot of sense without seeing an example, so let's create one.

Let's say that we have a structure that represents a baseball team. This structure has a property that represents the city the team is from and another property that represents the nickname of the team. The following code shows this structure:

```
struct BaseballTeam {
    let city: String
    let nickName: String
}
```

In this structure, if we need a way to retrieve the full name of the baseball team, including the city and nickName, we could probably create a method, as shown in the following example:

```
func fullname() -> String {
    return "\(city) \(nickName)"
}
```

This is how it can be done in most object-oriented programming languages. In our code that uses the BaseballTeam structure, we can then retrieve the city and nickname as properties with the dot notation and the full name from a method call. The following code shows how we can use both the city property and the fullname method:

```
var redsox = BaseballTeam(city: "Boston", nickName: "Red Sox")
let city = redsox.city
let fullname = redsox.fullname()
```

With dynamic member lookup, we can create a much cleaner interface. To use dynamic member lookup, the first thing we need to do is to add the @dynamicMemberLookup attribute when we define the BaseballTeam structure. This is shown in the following code, where we add some additional properties to our example:

```
@dynamicMemberLookup
struct BaseballTeam {
    let city: String
    let nickName: String
    let wins: Double
    let losses: Double
    let year: Int
}
```

Now, we will need to add the lookup to the `BaseballTeam` structure. This is done by implementing a dynamic member subscript like this: `subscript(dynamicMember: )`.

The following code shows how we would create a lookup to retrieve either the full name or the winning percentage for the `BaseballTeam` structure:

```
subscript(dynamicMember key: String) -> String {
    switch key {
    case "fullname":
        return "\(city) \(nickName)"
    case "percent":
        let per = wins/(wins+losses)
        return String(per)
    default:
        return "Unknown request"
    }
}
```

This code will use a `switch` statement, with the key that is passed in, to determine what information to return. With this code added to the `BaseballTeam` structure, we now have the ability to use the lookup:

```
var redsox = BaseballTeam(city: "Boston", nickName: "Red Sox", wins: 108,
                          losses: 54, year: 2018)
print("The \(redsox.fullname) won \(redsox.percent) of their games in
        \(redsox.year)")
```

Notice that we are able to access both `fullname` and `percent` from the instance of the `BaseballTeam` structure as if they were normal properties. This makes our code much cleaner and easier to read. However, there is one thing to keep in mind when using lookups like this: there is no way to control what keys are passed into the lookup.

In the previous example, we called `fullname` and `percent`; however, we could just as easily have called them using `flower` or `dog` with no warning from the compiler.

If you use dynamic member lookup, make sure you verify the key and handle any instances when something unexpected is sent, as we did in the previous example by using the default case of the `switch` statement.

Now that we have seen how to use dynamic member lookup, let's look at key paths.

Key paths

Key paths are powerful tools that allow us to access and manipulate properties in a type-safe, concise manner. They also provide a way to reference properties without using string literals, as we did with dynamic member lookup, which can be error-prone and difficult to maintain. Let's take a quick look at key paths to understand how we can use them.

Understanding key paths

A key path is a type-safe reference to a property of a type. It is represented by the KeyPath type, which is a generic type that includes two type parameters: the root type and the property type. They are defined using the \ syntax followed by the type and the property path. Let's look at a very basic example:

```
struct BasketballTeam {
    var city: String
    var nickName: String
}
let cityKeyPath = \BasketballTeam.city
```

In this example, we created a BasketballTeam structure and then created a cityKeyPath key path, which is an instance of KeyPathType that represents the city property of the BasketballTeam type. We would use this key path as follows:

```
var team = BasketballTeam(city: "Boston", nickName: "Celtics")
let teamCity = team[keyPath: cityKeyPath]
```

This example illustrates how we would access the value of a property with a key path using the subscript syntax. You can also set a property using key paths, as shown in the following example:

```
team[keyPath: cityKeyPath] = "Boston MA"
```

In this example, we use the key path to set a new value for the team's city. Key paths can also be used for nested properties, as shown here:

```
struct Season {
    let team: BasketballTeam
    let wins: Double
    let losses: Double
    let year: Int
}

let seasonTeamCityKeyPath = \Season.team.city
```

In this example, we create a Season structure that has a property named team of the BasketballTeam type. We then create a key path to the city property within the BasketballTeam type.

Static properties

Prior to Swift 6.1, Swift's key path syntax was limited to instance properties, as we saw in the previous section. However, with the introduction of **SE-0438**, developers can now reference static properties using key paths. This enables static properties to be accessed dynamically, similar to how instance properties are accessed with key paths.

Let's look at an example that highlights the difference between referencing an instance property and a static property:

```
struct Vehicle {
    static let maxSpeed = 100
    var currentSpeed = 0
}


let vehicleCurrentSpeed = \Vehicle.currentSpeed
let vehicleMaxSpeed = \Vehicle.Type.maxSpeed


print("Vehicle Max Speed = \(Vehicle.self[keyPath: vehicleMaxSpeed])")
```

In this example, \Vehicle.currentSpeed is a standard key path to an instance property while \Vehicle.Type.maxSpeed is a key path to a static property. Notice the use of .Type after the type name; this indicates that the key path is targeting a static property rather than an instance property.

Now, let's look at how we can use key paths within a function.

Key paths in functions

In the previous subsection, we mentioned that the KeyPath type is a generic type that includes two type parameters: the root type and the property type. This means that we can create generic functions that take a key path as an argument:

```
func getProperty<T, E>(of object: T, using keyPath: KeyPath<T, E>) -> E {
    return object[keyPath: keyPath]
}
```

This example defines a generic function named `getProperty` that retrieves a property from an object using the key path provided. We would use this function as follows:

```
let team = BasketballTeam(city: "Boston", nickName: "Celtics")
let city = getProperty(of: team, using: cityKeyPath)
```

In this code, the `city` constant would contain the string `"Boston"`.

One thing to keep in mind is that we do not need to define the key path prior to using it. For example, this code works as well:

```
let team = BasketballTeam(city: "Boston", nickName: "Celtics")
let city = getProperty(of: team, using: \.city)
```

Anywhere that we use a predefined key path, such as `cityKeyPath`, we can use `\.` in its place.

We can also use key paths with the `map` and `filter` functions.

The map and filter functions with key paths

Key paths can be used with higher-order functions such as `map` and `filter` to perform operations on collections in a clear and concise way. They enable us to reference properties of elements in a collection, which can be extremely useful for quickly accessing data without the need to write lengthy closures.

For the examples in this section, let's create a new `Person` type to use.

```
struct Person {
    let name: String
    let age: Int
}
```

The `Person` type contains two properties named `name` and `age`. Now, let's create an array that contains instances of the `Person` type that we can use for our `Map` and `Filter` functions:

```
let people = [
    Person(name: "Anna", age: 16),
    Person(name: "Bob", age: 40),
    Person(name: "Caroline", age: 27)
]
```

The people array that we created with this code contains three instances of the Person type. Now, let's look at how we can use the map function in combination with key paths with the people array we just created to get a list of names:

```
let names = people.map(\.name)
```

This code uses the map function with the \.name keypath to pull an array of names from the people array.

We could very easily do the same thing without key paths using a closure, as shown in the following code:

```
let names2 = people.map { $0.name }
```

Either of these two examples is fine to use, but the key path example is more concise and easier to read.

Now, let's see how we can use key paths with the filter function. The following code shows how we could pull a list of people who are 18 years old or older:

```
let adults = people.filter{ $0[keyPath: \.age] > 17 }
```

This code uses the filter function with the \.age key path to filter out any person who is under the age of 18.

Once again, we do not need to use key paths with the filter function and could very easily rewrite this code as shown here:

```
let _ = people.filter{ $0.age > 17 }
```

While using the map function with key paths produced code that was more concise and easier to read, the opposite is true for the filter function. Key paths are very powerful, and they enable us to access properties in a very flexible way; however, we should ensure that we use them properly and in a way that makes our code more concise and easier to understand.

Now, let's look at how we can use key paths and dynamic member lookups together.

Using key paths and dynamic member lookups together

While key paths and dynamic member lookups are very powerful tools on their own, each providing flexible ways to access properties, they become even more powerful when used together. Let's look at an example of how we can use them together.

We'll start with two types that, when combined, can be used to create a user profile:

```swift
struct Address {
    let street: String
    let city: String
    let state: String
    let zipCode: Int
}

struct User {
    let firstName: String
    let lastName: String
    let age: Int
    let address: Address
}
```

Here, we have defined two structures, Address and User, each representing different parts of a user's profile. Typically, we would create instances of the User type and access its properties like this:

```swift
let address = Address(street: "123 My Road", city: "Cupertino",
                      state: "Ca", zipCode: 95014)
let user = User(firstName: "Jon", lastName: "Hoffman", age: 56,
                address: address)

print(user.firstName)
print(user.address.street)
print(user.address.city)
```

Let's see how we can improve the way we access the properties by creating a user profile type that will take an instance of the User type. We'll use the @dynamicMemberLookup attribute when defining this type and create subscripts to access the properties:

```
@dynamicMemberLookup
struct UserProfile {
    let user: User

    subscript<T>(dynamicMember keyPath: KeyPath<User, T>) -> T {
        user[keyPath: keyPath]
    }

    subscript<T>(dynamicMember keyPath: KeyPath<Address, T>) -> T {
        user.address[keyPath: keyPath]
    }
}
```

The UserProfile type is a wrapper around an instance of the User type that uses the @dynamicMemberLookup attribute to simplify access to the properties of both the User and Address types.

The first subscript enables us to access the properties of the User type, while the second subscript enables us to access the properties of the Address type. Both of the subscripts take a key path to access the properties.

We can now create an instance of the UserProfile type and access the properties, as shown here:

```
let profile = UserProfile(user: user)
print(profile.firstName)
print(profile.lastName)
print(profile.age)

print(profile.street)
print(profile.city)
```

Notice that we do not need to use the .address path to access the properties of the embedded Address instance within the User instance. From this example, we can see how powerful combining key paths and dynamic member lookup is, significantly simplifying property access.

Summary

Dynamic member lookup allows properties to be resolved at runtime, creating a cleaner interface for accessing data. By using the `@dynamicMemberLookup` attribute and implementing a dynamic member subscript, properties can be accessed as if they were regular properties, making our code more concise and easier to read. However, caution is advised, as invalid keys won't trigger compiler warnings; therefore, it's important to put validation within the code.

Key paths provide a type-safe, concise method to access and manipulate properties without using string literals, thereby reducing errors. Represented by the `KeyPath` type and defined with the `\.` syntax, key paths can be used in functions to get and set property values. They also work with higher-order functions, enabling clean and easy-to-read operations on collections. We also saw how key paths and dynamic member lookup can be used together.

In the next chapter, we will look at how to use Grand Central Dispatch to manage concurrent and parallel operations.

Unlock this book's exclusive benefits now

Scan this QR code or go to `packtpub.com/unlock`, then search this book by name.

Note: Keep your purchase invoice ready before you start.

13

Grand Central Dispatch

Grand Central Dispatch (GCD) provides a simple, efficient way for developers to handle tasks. With GCD, we can assign tasks to different types of queues, which can run tasks one at a time (serial) or multiple tasks at once (concurrent). GCD is used particularly for managing concurrent and parallel operations effectively, simplifying the process of executing multiple tasks at the same time, which is essential for building high-performance and responsive applications. GCD's approach of using queues to organize and run tasks, rather than low-level threads, helps ensure that demanding operations, like data processing or network requests, don't adversely affect the user interface, keeping an app smooth and responsive.

In this chapter, we will cover the following topics:

- The difference between concurrency and parallelism
- How to use GCD

Let's start by looking at the difference between concurrency and parallelism.

Concurrency and parallelism

We often hear concurrency and parallelism used interchangeably; however, it is important to understand the difference between them.

Concurrency is the concept of multiple tasks starting, running, and completing within the same time period. This does not necessarily mean that the tasks are executed simultaneously. In fact, in order for tasks to be run simultaneously, our application needs to be running on a multicore or multiprocessor system. Concurrency allows us to share the processor or cores among multiple tasks, but a single core can only execute one task at any given time.

Parallelism is the concept of two or more tasks running simultaneously. Since each core of our processor can only execute one task at a time, the number of tasks executing simultaneously (in parallel) is limited to the number of cores within our processors and/or the number of processors that we have. As an example, if we have a four-core processor, then we are limited to running four tasks simultaneously. Today's processors can execute tasks so quickly that it may appear that larger tasks are executing simultaneously. However, within the system, the larger tasks are actually taking turns executing subtasks on the various cores.

In order to understand the difference between concurrency and parallelism, let's look at how a juggler juggles balls. If you watch a juggler, it seems they are catching and throwing multiple balls at any given time; however, a closer look reveals that they are, in fact, only catching and throwing one ball at a time. The other balls are in the air, waiting to be caught and thrown. If we want to be able to catch and throw multiple balls simultaneously, we need to have multiple jugglers.

Back when all of the processors were single-core, the only way to have a system that executed tasks simultaneously was to have multiple processors in the system. This also required specialized software to take advantage of the multiple processors. Today, just about every device has a processor that has multiple cores, and both iOS and macOS are designed to take advantage of these multiple cores to run tasks simultaneously.

Traditionally, the way applications added concurrency was to create multiple threads; however, this model does not scale well to an arbitrary number of cores. The biggest problem with using threads is that our applications may run on a variety of systems (and processors), and in order to optimize our code, we need to know how many cores/processors could be efficiently used at a given time, which is usually not known at the time of development.

To solve this problem, many operating systems, including iOS and macOS, started relying on asynchronous functions. These functions are often used to initiate tasks that could possibly take a long time to complete, such as making an HTTP request or writing data to disk. An asynchronous function typically starts a long-running task and then returns prior to the task's completion. Usually, this task runs in the background and uses a callback function (such as a closure in Swift) when the task completes.

These asynchronous functions work great for the tasks that the operating system provides them for, but what if we need to create our own asynchronous functions and do not want to manage the threads ourselves? For this, Apple provides several ways to perform concurrency and parallelism with Swift. In this chapter, we will look at GCD.

GCD

GCD provides dispatch queues to manage submitted tasks. These queues handle the submitted tasks and execute them in a First-In, First-Out (FIFO) order, ensuring that tasks start in the order they were submitted.

A task is any work that our application needs to perform. For example, tasks can perform simple calculations, read or write data to disk, make an HTTP request, or handle other necessary tasks. We define these tasks by placing the code inside a function or closure and adding it to a dispatch queue.

GCD provides three types of dispatch queues:

- **Serial queues**: Tasks in a serial queue (also known as a private queue) are executed one at a time in the order in which they were submitted. Each task starts only after the preceding task has been completed. Serial queues are often used to synchronize access to specific resources because they guarantee that no two tasks in the queue will run simultaneously. Therefore, if access to a specific resource is restricted to tasks within the serial queue, no two tasks will attempt to access the resource simultaneously or out of order.

- **Concurrent queues**: Tasks in a concurrent queue (also known as a global dispatch queue) execute concurrently, but they still start in the order in which they were added to the queue. The number of tasks that can run at any given time varies and depends on the system's current conditions and resources. The decision of when to start a task, beyond the order the tasks in which were submitted, is managed by GCD and is not something we can control within our application.

- **Main dispatch queue**: The main dispatch queue is a globally available serial queue that executes tasks on the application's main thread. Since tasks in the main dispatch queue run on the main thread, it is typically used to update the user interface after background processing has finished.

Dispatch queues offer several advantages over traditional threads. The primary benefit is that the system, rather than the application, handles the creation and management of threads. The system can dynamically scale the number of threads based on available resources and current conditions, allowing for more efficient thread management than manual handling.

Another advantage of dispatch queues is the ability to control the order in which tasks start. With serial queues, we not only manage the order of task execution but also ensure that a new task doesn't start until the previous one is complete. Implementing this with traditional threads can be cumbersome and fragile, but with dispatch queues, as we will see later in this chapter, it is quite easy.

Before we look at how to use dispatch queues, let's create a couple of functions that will help us demonstrate how the various types of queues work.

Creating functions for our queues

The first function we will create to demonstrate how GCD works will simply perform some basic calculations and then return a value. Here is the code for this function, which is named doCalc():

```
func doCalc() {
    let x = 100
    let y = x*x
    _ = y/x
}
```

The second function, which we will name performCalculation(), accepts two parameters. One is an integer named iterations, and the other is a string named tag. The performCalculation() function calls the doCalc() function repeatedly until it calls the function as many times as specified by the iterations parameter. We also use the CFAbsoluteTimeGetCurrent() function to calculate the elapsed time taken to perform all iterations. At the end of the function, we print the elapsed time along with the tag string to the console. This allows us to see when the function completes and how long it took to complete. Here is the code for this function:

```
func performCalculation(_ iterations: Int, tag: String) {
    let start = CFAbsoluteTimeGetCurrent()
    for _ in 0 ..< iterations {
        doCalc()
    }
    let end = CFAbsoluteTimeGetCurrent()
    print("time for \(tag):\(end-start)")
}
```

These functions will be used together to keep our queues busy so we can see how they work. Let's begin by looking at how we would create a dispatch queue.

Creating queues

To create a new dispatch queue, we use the DispatchQueue initializer. The following code shows how to do this:

```
let concurrentQueue = DispatchQueue(label: "cqueue.hoffman.jon",
                                    attributes: .concurrent)
let serialQueue = DispatchQueue(label: "squeue.hoffman.jon")
```

The first line would create a concurrent queue with a label of cqueue.hoffman.jon, while the second line would create a serial queue with a label of squeue.hoffman.jon. The DispatchQueue initializer takes the following parameters:

- label: This is a string that uniquely identifies the queue in debugging tools such as instruments and crash reports. It is recommended to use a reverse DNS naming convention, which is a naming convention used to create unique and descriptive identifiers based on domain names. This parameter is optional and can be nil.

- attributes: This specifies the type of queue to create. It can be DispatchQueue.Attributes.serial, DispatchQueue.Attributes.concurrent, or nil. If this parameter is nil, a serial queue is created. You can use .serial or .concurrent; if the attribute is left out, it defaults to .serial.

Now let's look at how we would create and use concurrent queues.

Creating and using a concurrent queue

A concurrent queue will execute tasks in FIFO order; however, the tasks will run concurrently and may finish in any order.

Let's see how to create and use a concurrent queue. The following code will create a concurrent queue and use the performCalculation function to test it:

```
let cqueue = DispatchQueue(label: "cqueue.hoffman.jon", attributes:.
concurrent)
let c : @Sendable () -> Void = {
    performCalculation(1000, tag: "async1")
}
cqueue.async(execute: c)
```

The first line will create a new dispatch queue that will be named cqueue. Next, a closure is created, which represents our task, and within the closure, the performCalculation() function is called, requesting that it runs through 1,000 iterations of the doCalc() function. Finally, we use the async(execute:) method of the queue to execute it. This code will execute the task in a concurrent dispatch queue, which is separate from the main thread.

While the preceding example works perfectly, we can actually shorten the code a little bit. The next example shows that a separate closure is not actually needed, and we can also submit the task to execute, as follows:

```
queue.async {
    performCalculation(1000, tag: "async2")
}
```

This shorthand version is how we usually submit small code blocks to our queues. If we have larger tasks, or tasks that we need to submit multiple times, we will generally want to create a closure and submit the closure to the queue, as we showed in the first example.

Let's see how a concurrent queue works by adding several items to the queue and looking at the order and time in which they return. The following code will add three tasks to the queue. Each task will call the performCalculation() function with various iteration counts.

> Remember that the performCalculation() function will execute the calculation routine continuously until it is executed the number of times defined by the iteration count. Therefore, the larger the iteration count, the longer it should take to execute.

Take a look at the following code:

```
let cqueue = DispatchQueue(label: "cqueue.hoffman.jon",
                           attributes:.concurrent)

cqueue.async {
    performCalculation(10_000_000, tag: "async1")
}
cqueue.async {
    performCalculation(1000, tag: "async2")
}
cqueue.async {
    performCalculation(100_000, tag: "async3")
}
```

Note that each of the functions is called with a different value in the tag parameter. Since the `performCalculation()` function prints out the tag variable with the elapsed time, we can see the order in which the tasks complete and the time they took to execute. If we execute the preceding code, we should see results similar to this:

```
time for async2:0.00031495094299316406
time for async3:0.026272058486938477
time for async1:2.1003830432891846
```

The elapsed time will vary from one run to the next and from system to system.

Since the queues function in a FIFO order, the task that had the tag async1 was executed first. However, as we can see from the results, it was the last task to finish. With this being a concurrent queue, if it is possible (if the system has the available resources), the blocks of code will execute concurrently. This is why tasks with the tags async2 and async3 were completed prior to the task that had the async1 tag, even though the execution of the async1 task began before the other two.

Now, let's see how a serial queue executes tasks.

Creating and using a serial queue

A serial queue operates differently from a concurrent queue. A serial queue will only execute one task at a time and will wait for one task to complete before starting the next. This queue, like the concurrent dispatch queue, follows the FIFO order.

The following line of code will create a serial queue that we will use for this section, with a label of squeue:

```
let squeue = DispatchQueue(label: "squeue.hoffman.jon")
```

Now, let's see how we would use this serial queue by using the `performCalculation()` function to perform some calculations:

```
let s : @Sendable () -> Void = {
    performCalculation(1000, tag: "async1")
}
squeue.async (execute: s)
```

In the preceding code, we created a closure, which represents our task, that calls the performCalculation() function, requesting that it runs through 1,000 iterations of the doCalc() function. Finally, the async(execute:) method of our queue is called to execute it. This code will execute the task in a serial dispatch queue, which is separate from the main thread. As we can see from this code, the serial queue is used just like the concurrent queue.

Just as with the concurrent queue, this code can be shortened. The following example shows how we would do this with a serial queue:

```
squeue.async {
    performCalculation(1000, tag: "async2")
}
```

Let's see how the serial queue works by adding several items to the queue and looking at the order in which they complete. The following code will add three tasks, which will call the performCalculation() function with various iteration counts to the queue:

```
let squeue = DispatchQueue(label: "squeue.hoffman.jon")
squeue.async {
    performCalculation(10_000_000, tag: "async1")
}
squeue.async {
    performCalculation(1000, tag: "async2")
}
squeue.async {
    performCalculation(100_000, tag: "async3")
}
```

Just as we did in the concurrent queue example, the performCalculation() function is called with various iteration counts and different values in the tag parameter. Since the performCalculation() function prints out the tag string with the elapsed time, we can see the order in which the tasks complete and the time it takes to execute. If we execute this code, we should see results similar to this:

```
time for async1:2.0900750160217285
time for async2:0.00024700164794921875
time for async3:0.020946025848388672
```

The elapsed time will vary from one run to the next and from system to system.

Unlike the concurrent queues, we can see that the tasks were completed in the same order that they were submitted, even though the sync2 and sync3 tasks took considerably less time to complete. This demonstrates that a serial queue only executes one task at a time and that the queue waits for each task to complete before starting the next one.

Executing code on the main queue function

The DispatchQueue.main.async(execute:) function will execute code on the application's main queue. We generally use this function when we want to update the UI or perform tasks that require changes to the UI.

The main queue is automatically created for the main thread when the application starts. This main queue is a serial queue; therefore, items in this queue are executed one at a time, in the order in which they were submitted. We will generally want to avoid using this queue unless we have a need to update the user interface from a background thread so that we do not block the main thread, which may make the user interface unresponsive.

The following example shows how we would use this function. For this code, it is assumed that an extension was used to add a method to the UIImage type that will resize the image:

```
let squeue = DispatchQueue(label: "squeue.hoffman.jon")
squeue.async {
    let resizedImage = image.resize(to: rect)
    DispatchQueue.main.async {
        picView.image = resizedImage
    }
}
```

In the code, a new serial queue is created and, in that queue, the image is resized. This is a good example of how to use a dispatch queue because we would not want to resize an image on the main queue as it could freeze the UI while the image is being resized.

Once the image is resized, the UIImageView is updated with the new image; however, all updates to the UI need to occur on the main thread. Therefore, the DispatchQueue.main.async function is used to perform this update.

In the previous examples, we used the async method to execute the code blocks. We could also use the sync method.

async versus sync methods with GCD

In the previous examples, the async method was used to execute the code blocks. When this method is used, the call will not block the current thread. This means that the method returns and the code block is executed asynchronously.

Rather than using the async method, the sync method could be used to execute the code blocks instead. The sync method will block the current thread, which means it will not return until the execution of the code has completed.

Generally, the async method is used, but there are use cases where the sync method is useful. These use cases are usually when we have a separate thread and we want that thread to wait for some work to finish.

Using asyncAfter

There will be times when we need to execute tasks after a delay. If we were using a threading model, we would need to create a new thread, perform some sort of delay or sleep function, and execute our task. With GCD, we can use the asyncAfter function.

The asyncAfter function will execute a block of code asynchronously after a given delay. This is very useful when we need to pause the execution of our code. The following code shows how we would use the asyncAfter function:

```
let queue2 = DispatchQueue(label: "squeue.hoffman.jon")
let delayInSeconds = 2.0
let pTime = DispatchTime.now() + delayInSeconds
queue2.asyncAfter(deadline: pTime) {
    print("Time's Up")
}
```

In this code, a serial dispatch queue is created. We then create an instance of the DispatchTime type and calculate the time to execute the block of code based on the current time. Finally, the asyncAfter function is used to execute the code block after the delay.

Now let's look at dispatch groups and how we can use them to coordinate the execution of multiple asynchronous operations and handle their results.

Dispatch groups

Dispatch groups provide a way to coordinate the execution of a set of tasks, allowing us to be notified when all tasks in the group have completed. This is useful when we need to perform multiple independent tasks concurrently and wait for all of them to finish.

Take a look at the following code to see how this works:

```
let queue = DispatchQueue(label: "cqueue.hoffman.jon",
                          attributes:.concurrent)

let dispatchGroup = DispatchGroup()

dispatchGroup.enter()
queue.async {
    print("async1 started")
    performCalculation(10_000, tag: "async1")
    print("aync1 completed")
    dispatchGroup.leave()
}

dispatchGroup.enter()
queue.async {
    print("async2 started")
    performCalculation(1_000, tag: "async2")
    print("async2 completed")
    dispatchGroup.leave()
}

dispatchGroup.notify(queue: DispatchQueue.main) {
    print("All tasks are complete")
}
```

This code starts by creating a concurrent queue, just as we have done in previous examples in this chapter. We then create a dispatch group. Notice that before each task or queue is started, the dispatchGroup.enter() method is called and, at the completion of each task or queue, the dispatchGroup.leave() method is called. This is to ensure that the dispatch group is aware of each task's lifecycle.

The dispatchGroup.notify(queue:) method is called once all of the tasks or queues have called their leave() methods and the notify method is called on the main queue.

The output of this code should look similar to this:

```
async2 started
async1 started
time for async1:0.00032806396484375
aync1 completed
time for async2:0.00229799747467041
async2 completed
All tasks are complete
```

Now that we have seen how dispatch groups provide a way to coordinate the execution of a set of tasks, let's take a look at DispatchWorkItem.

DispatchWorkItem

With DispatchWorkItem, we can encapsulate a block of code into a work item that can be executed asynchronously. This also provides us with greater control over the task's execution, enabling us to cancel it, attach completion handlers, and establish task dependencies. Let's see how this works:

```
let workItem = DispatchWorkItem {
    for i in 1...9 {

        if workItem.isCancelled {
            print("workItem was cancelled")
            break
        }
        print("Executing \(i)")
        Thread.sleep(forTimeInterval: 1)

    }
}
```

This code begins by creating an instance of DispatchWorkItem named workItem. This work item executes a loop from 1 to 9, printing a message at each iteration and pausing for a second in each iteration.

Within this loop, we also call isCancelled to check whether the item has been canceled. If it has, we print a message and exit the loop. Before exiting the task, we would typically perform any necessary cleanup.

Now, let's see how we would use this work item:

```
workItem.notify(queue: DispatchQueue.main) {
    print("workItem has completed")
}


DispatchQueue.global(qos: .background).async(execute: workItem)


let delayInSeconds = 4.0


DispatchQueue.global().asyncAfter(deadline: .now() + delayInSeconds) {
    print("Cancelling workItem")
    workItem.cancel()
}
```

We begin by adding a completion handler using the notify() method. This method accepts a closure that will be executed when the work item finishes or is canceled. Next, we schedule the work item for asynchronous execution in the background queue. Finally, we use asyncAfter, which we saw earlier in the chapter, to cancel the task after 4 seconds.

Now let's look at DispatchTime and see how it can be used to delay the execution of a task.

DispatchTime

DispatchTime represents a point in time relative to the system's clock. It is commonly used when we want to schedule tasks for execution after a specific delay. Let's see how we would use it:

```
let delayInSeconds = 4.0


let delayTime = DispatchTime.now() + delayInSeconds


DispatchQueue.main.asyncAfter(deadline: delayTime) {
    print("After a \(delayInSeconds) second delay.")
}
```

Here, we begin by defining the duration of our delay to be four seconds. Next, we define the delayTime constant using the static now() method of the DispatchTime structure to obtain the current system time, adding the delay to this value. Finally, we employ the asyncAfter function, specifying the deadline parameter with the value of the delayTime constant. After the four-second delay, the message is printed out.

While `DispatchTime` is based on the system clock and pauses when the system goes to sleep, `DispatchWallTime` tracks the actual passage of time continuously. Let's take a look at this.

DispatchWallTime

`DispatchWallTime` is useful when we need to schedule work based on the actual passage of time, regardless of whether the device goes to sleep or not. `DispatchWallTime` works in a similar way as `DispatchTime`. Let's see how to use it:

```
let delayInSeconds = 4.0

let delayTime = DispatchWallTime.now() + .seconds(5)

DispatchQueue.main.asyncAfter(wallDeadline: delayTime) {
    print("After \(delayInSeconds) second delay.")
}
```

We begin by defining the duration of our delay to be four seconds. Next, we define the `delayTime` constant using the static now() method of the `DispatchWallTime` structure to obtain the current time, adding the delay to this value. Finally, we employ the asyncAfter function, specifying the wallDeadline parameter with the value of the `delayTime` constant. After the four-second delay, the message is printed out.

Next, let's take a look at how to synchronize the execution of tasks by using barriers.

Barriers

Barriers provide a way to synchronize the execution of tasks, ensuring that a particular task completes before other tasks are executed. This is very useful when we are working with concurrent queues.

Let's see how we would use barriers through the following code:

```
let queue = DispatchQueue(label: "cqueue.hoffman.jon", attributes:.
concurrent)

queue.async {
    print("async1 started")
    performCalculation(30_000, tag: "async1")
    print("aync1 completed")
}
```

```
queue.async {
    print("async2 started")
    performCalculation(10_000, tag: "async2")
    print("async2 completed")
}

queue.async(flags: .barrier) {
    print("async3 started")
    performCalculation(100_000, tag: "async3")
    print("async3 completed")
}

queue.async {
    print("async4 started")
    performCalculation(100, tag: "async4")
    print("async4 completed")
}
```

This code starts by creating a concurrent queue just as we have done in previous examples in this chapter. The async function is then called four times to add four concurrent tasks. Note that the third time the async function is called, we use the .barrier flag.

The .barrier flag acts as a synchronization point, which ensures that the "barrier" task (async3, in our code) is executed only after all previously submitted tasks in this queue have completed. Additionally, any tasks submitted after this task will not start until the barrier task has finished executing.

Now let's look at how we can synchronize access to certain resources using dispatch semaphores.

Dispatch semaphores

Dispatch semaphores are used to synchronize access to a resource by multiple threads. They help manage concurrent execution by limiting the number of threads that can access a particular resource simultaneously, preventing race conditions. (We look at race conditions in the next chapter.)

A semaphore uses a counter to keep track of the number of available permits, which represent permission to access a shared resource. Threads or tasks must acquire a permit before accessing the resource. If a permit is available, the counter decreases, and the thread or task proceeds. If no permits are available, the thread or task waits until another thread releases a permit.

Let's look at an example of how we would use a dispatch semaphore:

```
var cnt = 0

let semaphore = DispatchSemaphore(value: 1)
let queue = DispatchQueue(label: "cqueue.hoffman.jon", attributes:.
concurrent)

func accessSharedResource(taskNumber: Int) {
    semaphore.wait()


    print("Task \(taskNumber) is accessing the resource (\(cnt))")
    sleep(2) // Simulates a delay

    cnt += taskNumber
    print("Task \(taskNumber) is releasing the resource (\(cnt))")
    semaphore.signal()
}
```

This code begins by creating a variable named cnt, which serves as a shared resource in this example. Next, a dispatch semaphore is defined with an initial value of 1. This value indicates the number of simultaneous accesses allowed to the shared resource; in this example, we restrict it to just one thread at any given time. Following this, a concurrent queue is established.

The accessSharedResource() function is designed to increment the cnt variable by the taskNumber parameter. Since this function could be executed across multiple threads and may operate asynchronously, it is necessary to regulate access to the cnt variable to ensure that only one thread can modify it at any moment.

In order to regulate access to the cnt variable within the accessSharedResource() function, the function starts by calling the wait() method on the DispatchSemaphore instance. This method attempts to decrement the semaphore's counter, and if the counter is greater than zero, it proceeds. If the counter is zero, the task is blocked until a permit is available.

When the task can proceed, the cnt variable is incremented by the taskNumber parameter after a simulated delay. Once the cnt variable has been incremented, the signal() function is called on the DispatchQueue instance, which releases the permit and increments the semaphore's counter.

The following code creates five concurrent tasks that will attempt to access the accessSharedResource() function concurrently:

```
for i in 1...5 {
    queue.async {
        print("submitting \(i)")
        accessSharedResource(taskNumber: i)
    }
}
```

When we run this code, we should see output similar to this:

```
submitting 1
submitting 3
submitting 2
submitting 4
submitting 5
Task 1 is accessing the resource (0)
Task 1 is releasing the resource (1)
Task 3 is accessing the resource (1)
Task 3 is releasing the resource (4)
Task 2 is accessing the resource (4)
Task 2 is releasing the resource (6)
Task 4 is accessing the resource (6)
Task 4 is releasing the resource (10)
Task 5 is accessing the resource (10)
Task 5 is releasing the resource (15)
```

Note that semaphores only regulate the number of tasks that can access a resource simultaneously; they do not determine the order in which access is granted, as shown in the preceding output, where Task 3 accessed the resource prior to Task 2.

Summary

We started this chapter by explaining the concepts of concurrency and parallelism, which are often confused but have distinct meanings in software development. Concurrency involves managing multiple tasks in overlapping time periods, without them necessarily running simultaneously, while parallelism is running multiple tasks simultaneously.

Next, we discussed how these concepts apply in software development using Swift's Grand Central Dispatch. GCD organizes tasks into queues, which can be either serial or concurrent. Serial queues handle one task at a time in the order they are added, ideal for tasks that need to be done in sequence. Concurrent queues allow tasks to run simultaneously, speeding up operations that don't depend on each other.

Choosing between serial and concurrent queues gives us flexibility. Serial queues prevent data issues by ensuring only one task accesses shared resources at a time. Concurrent queues maximize the use of multi-core processors by performing multiple tasks at once, reducing the time needed for large volumes of work.

Additionally, GCD provides functions for managing tasks related to the user interface and scheduling tasks to run later. For example, `DispatchQueue.main.async` is used to safely update the user interface on the main thread. The `asyncAfter` function delays tasks, allowing precise control over the timing of operations. These tools make GCD a powerful component for modern Swift developers, simplifying asynchronous task management and improving application responsiveness.

In the next chapter, we will look at asynchronous operations with Swift.

Unlock this book's exclusive benefits now

Scan this QR code or go to packtpub.com/unlock, then search this book by name.

Note: Keep your purchase invoice ready before you start.

14

Structured Concurrency

Concurrency is an important aspect of modern programming. It enables applications to perform multiple operations simultaneously, improving efficiency and responsiveness. In Swift, concurrency has become easier to work with thanks to new features such as async/await, tasks, and task groups. These tools enable developers to write code that can handle multiple operations simultaneously, without affecting code readability or introducing errors such as race conditions and deadlocks, which are common issues when dealing with concurrent tasks.

With these new concurrency tools, we can manage tasks such as downloading data or reading/writing files more efficiently and seamlessly. async/await also makes writing asynchronous code more intuitive, while tasks help organize these operations better.

In this chapter, we will cover the following topics:

- Data races, and how Swift 6 aims to prevent them
- How to use async and await to make asynchronous calls
- Tasks, and how to use and group them
- Actors, and how they can make our concurrent code safer
- Strict concurrency checking, and how to opt in to this

Let's start off by looking at data race conditions and Swift 6.

Data race conditions and Swift 6

One of the major goals of Swift 6 was to ensure data race safety, and it introduces key changes designed to help concurrent code run predictably and to avoid crashes.

A data race happens when multiple threads simultaneously access the same piece of mutable data at the same time, where at least one of these modifies the data. This can lead to unpredictable behavior and even crashes, since the order in which the data is accessed depends on the random timing of these concurrent accesses.

Let's take a look at what a data race condition is through the following code:

```swift
var count = 0
let queue = DispatchQueue.global()

func incrementCount() {
    for _ in 0..<1000 {
      queue.async {
        count += 1
          print(count)
      }
    }
}

incrementCount()

print("Final:  \(count)")
```

When this code is run, one of the first things that would be noticed is that the output from within the incrementCount() function would be out of order. The end of the output may look something like this:

```
938
876
875
941
942
```

Notice that the final number, which should be 999, is not the final number in the output. If we scrolled through the output, we would notice that the output from the last line of code is probably somewhere in the middle, and the count value that was output would be a number probably below 500. The reason for this is that multiple current threads are reading and modifying the data based on the timing of each thread's execution.

> 💡 **Quick tip**: Enhance your coding experience with the **AI Code Explainer** and **Quick Copy** features. Open this book in the next-gen Packt Reader. Click the **Copy** button
>
> **(1)** to quickly copy code into your coding environment, or click the **Explain** button
>
> **(2)** to get the AI assistant to explain a block of code to you.
>
> ```
> function calculate(a, b) {                                    Copy    Explain
>   return {sum: a + b};                                         1        2
> };
> ```
>
> 📖 **The next-gen Packt Reader** is included for free with the purchase of this book. Scan the QR code OR go to packtpub.com/unlock, then use the search bar to find this book by name. Double-check the edition shown to make sure you get the right one.

Swift 6 introduces a number of features to help prevent these types of data race conditions at compile time, including the use of:

- The async and await keywords
- Tasks and task groups
- Actors
- Sendable types

At the core of these is **strict concurrency checking**, which is an opt-in feature. When this is enabled, the compiler will analyze the code for potential data race conditions and issue warnings or errors if any are detected. This approach helps us identify and fix any potential issues during the development process. We will look at this in more detail later in the chapter.

To begin our exploration of Swift 6's approach to preventing data race conditions, let's now look at asynchronous functions in Swift.

Asynchronous functions

Managing asynchronous tasks efficiently is critical to ensure a smooth user experience in our applications. Asynchronous operations, such as network requests, file I/O, or heavy computations, can cause an application to become unresponsive and appear to hang if not managed properly. Swift's async and await keywords provide a robust and intuitive way to manage these operations.

The async keyword is used to mark asynchronous functions, indicating that they may pause execution to wait for the completion of another asynchronous task. This allows for operations such as network calls to be performed without blocking the main thread, ensuring that the user interface remains responsive.

The await keyword is used to call these asynchronous functions, indicating that the function's execution may be suspended until the awaited task is complete.

> RunLoop.main.run() is used in the downloadable code samples for this chapter to keep the application from exiting before the code is finished. This should not be used in production applications because there is no async method defined for RunLoop.

Let's take a look at how we would use async and await.

async and await

Including async in a function definition indicates that the function may pause its execution to wait for another asynchronous operation to finish.

The following code shows two examples of how we would use the async keyword within a function definition:

```
func retrieveData() async -> String {}


func loadContent() async {}
```

The first function definition demonstrates how to create an asynchronous function that returns information, while the second one illustrates how to define an asynchronous function that does not return a value.

Now, let's see how await is used with asynchronous functions. For our example, the retrieveData() function will use the Task.sleep(nanoseconds:) function to simulate a delay in retrieving data from an online service. This is the code for the retrieveData() function:

```
func retrieveData() async -> String {
    print("Retrieving data")
    try! await Task.sleep(nanoseconds: 2_000_000_000)
    return "Data Retrieved"
}
```

This function starts by indicating that it has begun retrieving data and then waits for two seconds. After the delay, it returns a message confirming that the data has been retrieved.

The loadContent() function will simulate displaying the data that is retrieved with the retrieveData() function. This is the code for the loadContent() function:

```
func loadContent() async {
    let data = await retrieveData()
    print("Data: \(data)")
}
```

When these functions are run, we will see the Retrieving Data message, followed by a two-second delay, and then we will see the Data: Data Retrieved message.

We may think that when the await keyword is used with asynchronous functions, it spins the function off to another thread; however, it actually pauses the execution of the current function until the awaited operation is completed. Pausing the execution of the function enables other work to proceed until the awaited operation has finished.

Calling multiple asynchronous functions

When we need to call multiple asynchronous functions, we may need to coordinate the tasks to ensure that they are completed in a predictable manner. This will help prevent common concurrency issues such as race conditions and deadlocks. Let's look at how this can be done.

For this example, let's create two asynchronous functions called `retrieveUserData()` and `retrieveImageData()`. Here is the code for these functions:

```swift
func retrieveUserData() async -> String {
    print("Retrieving user data")
    try? await Task.sleep(nanoseconds: 2_000_000_000)
    return "User Data Retrieved"
}


func retrieveImageData() async -> String {
    print("Retrieving image data")
    try? await Task.sleep(nanoseconds: 4_000_000_000)
    return "Image Data Retrieved"
}
```

These functions use the `Task.sleep(nanoseconds:)` function to simulate the delay in retrieving information from a remote source.

Now, let's assume we would want to wait for both of these calls to finish before updating the UI. We can coordinate the calls, ensuring that they both finish, with the following code:

```swift
async let userData = retrieveUserData()
async let imageData = retrieveImageData()


let results = await (userData, imageData)
print("User Data: \(results.0) \nImage Data: \(results.1)")
```

Notice that, in the first two lines of this code, we use the `async let` syntax to initiate the asynchronous calls to the `retrieveUserData()` and `retrieveImageData()` functions. This enables them to run concurrently. The `await` keyword is then used to wait for both of these asynchronous operations to complete. By awaiting the tuple `(userData, imageData)`, we ensure that the application only proceeds once both functions finish and return.

A core concept in Swift's concurrency model is tasks. Let's now take a look at them.

Tasks

A task represents a unit of work that can be run asynchronously and provides an alternative to dispatch queues, which we looked at in the previous chapter. Tasks enable functions to run concurrently without blocking the main thread, similar to what we just saw with the `async` and `await` keywords; however, tasks give us additional control.

To create a task, we use the task initializer, which takes a closure containing the code to execute asynchronously. This is the syntax to create a task:

```
Task {
    //asynchronous code here
}
```

Note that using the task API does not necessarily create a new thread. Instead, tasks leverage Swift's concurrency model, which uses cooperative thread management. When a task is created using `async let` or `await`, Swift manages the execution of these tasks within the same thread or across different threads managed by the system. It doesn't guarantee the creation of a new thread for each new task. When a task is created using `Task { }`, it runs the provided code asynchronously. The execution context is managed by Swift's runtime and also may not involve creating a new thread.

The following code shows how we would create a task that would use the `retrieveUserData()` function we created previously:

```
Task {
    let data = await retrieveUserData()
    print("Data: \(data)")
}
```

From this code, we can see that we call the `retrieveUserData()` function just as we previously did, but it is now within a task. This may seem unnecessary, and for a very basic asynchronous call, which would be correct; however, tasks give us a lot of additional control over our asynchronous calls. For example, we can detach them, which means they are no longer bound to their parent task.

Detaching tasks

Detaching a task creates a new, independent task that is not tied to the parent task. We would detach a task when we want to perform work independently of the current task. We can detach a task like this:

```
Task.detached {
    let data = await retrieveUserData()
    print("Data: \(data)")
}
```

In this code, we use the detached method of the task to detach the task.

Once again, detached tasks do not necessarily create new threads. Swift's concurrency model uses a cooperative thread pool, which is managed by the system. This model optimizes task execution across available threads, ensuring that the tasks and your application run efficiently, no matter what type of system it is running on, without the need for explicitly creating and managing threads.

Detached tasks do not inherit the context or actor from the scope where it was created. What this means is that since the detached task operates outside of the current context, a detached task cannot access or modify the properties of the actor or any shared data in the calling context. It creates a separate, independent task, making it useful for tasks that don't need access to the actor's shared state.

Canceling tasks

Another feature of Swift's tasks is the ability to cancel them, but this is done through cooperative cancellation. What this means is that, while a task can be canceled, it has the freedom to continue running until it reaches a suitable point to stop. This approach ensures that tasks can perform the necessary cleanup and exit gracefully, which helps in properly releasing resources and maintaining a consistent state for the application.

Within the task, we can use the Task.isCancelled property to periodically check whether a cancellation request has been made. Based on this check, the task can decide to perform any necessary cleanup operations and/or exit early.

Let's see an example of canceling a task, first without using the Task.isCancelled property:

```
let task = Task {
    for i in 0..<10 {
        print("Loop \(i)")
```

```
        let data = await retrieveUserData()
    }
    print("Task completed successfully.")
}
```

In this code, we create a new task and, within the task, a for loop is used to call the retrieveUseData() function, which we created earlier in this chapter, 10 times. The retrieveUseData() function calls the Task.sleep(nanoseconds:) function to pause for two seconds. The following code shows this:

```
try? await Task.sleep(nanoseconds: 6_000_000_000)
task.cancel()

await task.value
print("Task finished.")
```

Outside of the task, we use the Task.sleep(nanoseconds:) function to pause for six seconds and then call the cancel() function on the task to cancel it.

When this code is run, we can see that the task completes the first three iterations with a two-second pause between each. However, after these first three iterations, the remaining seven iterations occur almost instantaneously. This happens because the task cancels all its child tasks (the Task. sleep(nanoseconds:) calls) and attempts to clean up properly before exiting.

Now let's see what happens if we check to see whether the task is canceled at each iteration. If the task is canceled, we can perform any necessary cleanup and then exit the task. The following code illustrates this:

```
let task = Task {
    for i in 0..<10 {
        if Task.isCancelled {
            print("Task was cancelled, cleaning up")
            return
        }

        print("Loop \(i)")
        let data = await retrieveUserData()
    }
    print("Task completed successfully.")
}
```

```
try? await Task.sleep(nanoseconds: 6_000_000_000)
task.cancel()

await task.value
print("Task finished.")
```

Note that the only change in this code is the `Task.isCancelled` check. When this code is run, we will see that after the third iteration of the `for` loop, the `Task was cancelled, cleaning up` message appears, and the task exits without any further iterations of the `for` loop being run.

When a task is canceled, we can throw the `CancellationError()` error to indicate that the task was canceled, enabling our application to perform any cleanup outside of the task. To demonstrate this, let's refactor the code and place the code that was in the task within a function. The function will look like this:

```
func testCancelTask() async throws {
    for i in 0..<10 {
        if Task.isCancelled {
            print("Task was cancelled, cleaning up")
            throw CancellationError()
        }

        print("Loop \(i)")
        let _ = await retrieveUserData()
    }
}
```

Notice that the function is defined as throwing an error, and instead of using the `return` command to exit out of the task, we throw the `CancellationError()` error instead.

We could now call this function like this:

```
let task = Task {
    do {
        try await testCancelTask()
    } catch is CancellationError {
        print("Caught a cancellation error")
    } catch {
        print("Caught an unexpected error: \(error)")
    }
```

```
    }
    try? await Task.sleep(nanoseconds: 6_000_000_000)
    task.cancel()
```

In this code, we call the `testCancelTask()` function with the `do/catch` block and then catch the `CancellationError` error. Now, if we need to do any cleanup outside of the task, we can do it within the `catch` block. Now let's see how we can request that a task start immediately.

Starting a task synchronously

When we create a task, it is queued to run at the next opportunity; however, with Swift 6.2, we have the ability to request that a task start immediately. To do this, we use the immediate function to request that the task be run immediately. Let's look at an example of this:

```
Task.immediate {
    let data = await retrieveUserData()
    print("Data: \(data)")
}
```

When we request that the task runs immediately, that does not guarantee that it will run immediately, but it will if possible. Now let's look at how we can name tasks.

Task naming

Swift 6.2 introduced another useful enhancement to tasks, which is the ability to assign names to them. This feature enables us to identify tasks by name, which is especially valuable for debugging and monitoring. You can name tasks when creating them using `Task.init()` or `Task.detached()`, and when creating child tasks in task groups, using `addTask()` or `addTaskUnlessCancelled()`, which we'll explore later in this chapter. Now let's look at how we can name a task.

```
let task = Task(name: "Task1") {
    print("Current task \(Task.name ?? "Unknown") has started")
}
```

In this code, we used the `Task(name:)` initializer to create our task and then used the `name` property of the current task to print out the name of the task to the console. Now that we have seen what tasks can do for us, let's look at task groups.

Task groups

A task group enables us to create and manage a collection of related tasks that can be run concurrently. They are particularly useful when we have multiple pieces of work that can be done in parallel but need to be grouped together, with the results collectively managed.

Before we look at how to create task groups, let's create a new `retrieveUserData()` function that takes a single parameter and has a random delay:

```
func retrieveUserData(_ forUser: String) async -> String {
    print("Retrieving user data for \(forUser)")
    try? await Task.sleep(for: .seconds(Double.random(in: 1...6)))
    return "User Data Retrieved for \(forUser)"
}
```

We will use this new `retrieveUserData()` function to illustrate how task groups work. The following code creates a task group and uses this new function:

```
func taskGroup() async -> [String] {
    return await withTaskGroup { group in
        group.addTask {
            await retrieveUserData("Jon")
        }

        group.addTask {
            await retrieveUserData("Heidi")
        }

        group.addTask {
            await retrieveUserData("Kailey")
        }

        group.addTask {
            await retrieveUserData("Kai")
        }

        var data = [String]()
        for await string in group {
            data.append(string)
```

```
        }
        return data
    }
}
```

In this function, we create a task group using the await `withTaskGroup()` function. If the task group threw an error, we would use the `withThrowingTaskGroup()` function. Within the definition, we define that the results that the tasks produce are of the `String` type. If the tasks do not produce any results, we use `Void.self` instead of `String.self` for the type.

We then add four tasks to the task group, each asynchronously calling the `retrieveUserData()` function for different users ("Jon", "Heidi", "Kailey", and "Kai"). It's important to note that since each task is called asynchronously, there is no guarantee that the tasks will run in the order they were added.

The `for await string in group {}` statement will wait for all tasks in the group to finish before iterating over the results. Within this loop, we add the results of the tasks to an array and return this array.

We can call this function like this:

```
let results = await taskGroup()
for result in results {
    print("Result: \(result)")
}
```

When we run this code, we will see results similar to this:

```
Retrieving user data for Heidi
Retrieving user data for Kai
Retrieving user data for Kailey
Retrieving user data for Jon
Result: User Data Retrieved for Kai
Result: User Data Retrieved for Heidi
Result: User Data Retrieved for Kailey
Result: User Data Retrieved for Jon
```

You will notice that the order in which the tasks are executed and completed may not match the order in which they were initiated. This is an important consideration when using task groups.

Next, let's look at the `Actor` type and how it can help us avoid common concurrency issues such as data race conditions.

Actors

Actors are a powerful concurrency feature that helps to manage state safely in concurrent environments. They are designed to handle mutable state while preventing common concurrency issues. By using actors, we ensure that only one thread can access the actor's mutable state at a time. This can greatly simplify the development of concurrent code.

> A useful way to think about actors is that they pass messages to the instance, rather than references to the instance.

Actors are reference types, similar to classes, so they can be used to share state. They can also have properties, methods, initializers, and subscripts, and can conform to protocols and be generic. However, unlike classes, actors do not support inheritance.

Let's look at an example of an actor and see the problem they are designed to solve. For this example, we will create a very basic BankAccount type as follows:

```swift
actor BankAccount {
    private var balance: Double

    init(_ balance: Double) {
        self.balance = balance
    }

    func deposit(amount: Double) {
        balance += amount
    }

    func withdraw(amount: Double) -> Bool {
        if balance >= amount {
            balance -= amount
            return true
        } else {
            return false
        }
    }
}
```

```
    func getBalance() -> Double {
        return balance
    }
}
```

The BankAccount type is defined as an actor type. A type like this should not be defined using a value type, such as a struct, because the account's state (balance) may need to be shared across different parts of our code. Defining it as a class also wouldn't be appropriate, as it could lead to issues with data race conditions. This is where actors come in, providing a safe way to manage mutable state in concurrent environments.

Within the BankAccount type, we define a single private property called balance, which holds the account's current balance. Without a safe way to manage the state of the balance property in a concurrent environment, multiple threads may attempt to update the balance at the same time, leading to various issues.

Actor isolation is the concept that an actor's mutable state can only be accessed within the actor itself. By making the balance property private (only accessible through the methods defined within the actor), we ensure that the balance property is not modified concurrently from different threads.

We use actors similarly to other types, with the distinction that their methods are asynchronous, even if not explicitly marked as such. The following code demonstrates how to use the BankAccount actor:

```
let account = BankAccount(5000)

Let _ = await account.withdraw(amount: 100)
print("New Balance \(await account.getBalance())")
```

In this code, we create an instance of the BankAccount type, withdraw $100, and then show the new balance. Notice that the await keyword is used when calling the withdraw() and getBalance() functions.

Global actors

While regular actors are great for isolating state within a specific instance, as we saw with the previous BankAccount example, sometimes we need to coordinate shared functionality across multiple parts of our code. That's where global actors can help.

A global actor is a singleton-like actor that defines a shared execution context. Any function, property, or type marked with the global actor attribute will always run on that actor's isolated context. This ensures consistent, thread-safe behavior across our application or module. This is particularly useful for shared services or resources.

Let's look at a simple example. Suppose we want to build a basic logging system that can be safely used from anywhere in our application or module. Where logs may come from multiple places at once, possibly even from concurrent tasks, we would need to ensure that those logs are handled in a thread-safe way. Here's how we could do this using a custom global actor:

```
@globalActor
actor LoggerActor {
    static let shared = LoggerActor()
}
```

The `globalActor` attribute does have a single requirement, which is a static property named `shared`. Now we can use `@LoggerActor` to mark any function, property, or type that should run in this shared context:

```
@LoggerActor
struct Logger {
    static func log(_ message: String) {
        print("[\(Date())] \(message)")
    }
}
```

Because the `log(_:)` method is isolated to the `LoggerActor` global actor, by marking the `Logger` type itself, we ensure it will never be executed concurrently with another `LoggerActor` call. That eliminates race conditions even if we call the log method from multiple tasks at once.

For the final piece of Swift's concurrency model, let's look at sendable types.

Sendable types

A sendable type in Swift is a type that conforms to the `Sendable` protocol, which means that instances of the type can be safely transferred between different threads and tasks.

There are many things in Swift that are inherently safe to send across threads and internally already conform to the `Sendable` protocol. Some of these are:

- All of Swift's core value types, such as `Int`, `String`, and similar types

- Optionals whose wrapped data is a value type
- Standard collections that contain value types
- Tuples whose elements are all value types

For custom types to conform to the Sendable protocol, the following rules apply:

- Actors automatically conform to the Sendable protocol.
- Custom value types such as structs and enumerations automatically conform to the Sendable protocol if they contain only values that also conform to the Sendable protocol.
- Reference types, such as classes, can conform to the Sendable protocol if:

  - They do not inherit from another reference type.
  - All properties are constants (let) and conform to the Sendable protocol.
  - A class is marked as final to prevent further inheritance.

These rules ensure that only types that are safe to use concurrently have the ability to conform to the Sendable protocol, avoiding common concurrency issues.

Now let's look at an example by creating a Transaction type that conforms to the Sendable protocol:

```
struct Transaction: Sendable {
    let id: Int
    let amount: Double
    let description: String
}
```

This Transaction type can conform to the Sendable protocol because the three properties already conform to Sendable. This indicates that the Transaction type is safe to use across various threads and tasks. We can now use this type within an actor, for example, like this:

```
actor BankAccount1 {
    private var transactions = [Transaction]()

    func addTransaction(_ transaction: Transaction) {
        transactions.append(transaction)
    }
}
```

In the BankAccount1 actor, we define an array that contains a list of transactions for the account, and then a function that can be used to add a transaction. The sendable type is a crucial part of Swift's concurrency model because it provides a guarantee that the data is safe to use across threads and tasks.

There are situations where the compiler cannot verify that a type is safe for concurrent use; however, we would still like it to conform to Sendable. This often happens with classes that use internal synchronization mechanisms, such as locks, or when a type contains non-Sendable members. In these cases, if we are confident that the type is thread-safe, we can use the @unchecked Sendable attribute.

Let's see how to apply the @unchecked Sendable attribute to a class that manages its own synchronization:

```
class Counter: @unchecked Sendable {
    private var count = 0
    private let lock = NSLock()

    func increment() {
        lock.lock()
        defer { lock.unlock() }
        count += 1
    }

    func getCount() -> Int {
        lock.lock()
        defer { lock.unlock() }
        return count
    }
}
```

In this example, the Counter class uses an NSLock instance to synchronize access to the count variable. This ensures that only one thread can read or modify the count variable at a time, providing thread safety. However, the compiler cannot automatically verify this because it doesn't analyze the behavior of locks or other synchronization mechanisms. By declaring the class as @unchecked Sendable, we explicitly tell the compiler that we have manually ensured thread safety. This allows instances of Counter to be shared across concurrent contexts without generating compiler warnings or errors.

It's important to consider that using @unchecked Sendable places the responsibility on us to ensure that the type is truly thread-safe, as the compiler will not enforce or check the thread safety. Therefore, it's crucial to thoroughly test the type in concurrent scenarios to verify that no data races or concurrency issues exist.

We will often run into legacy APIs that still use completion handlers. Thankfully, the Swift developers thought of this and gave us a clean way to bridge the gap using continuations.

Adapting completion handlers to async/await

With the introduction of Swift's structured concurrency model in Swift 5.5, asynchronous programming has become cleaner, easier to read, and more manageable. Instead of having to rely on completion handlers, we can now write asynchronous code that looks and behaves more like synchronous code using async/await, as we have seen throughout this chapter. However, many existing APIs, especially legacy ones, still rely on completion handlers. In this section, we'll look at how to bridge the gap between those legacy patterns and Swift's modern concurrency model using the withCheckedContinuation and withUnsafeContinuation functions.

Let's say you're using a utility that performs an API call using a completion handler. For this example, we will use the following code:

```
func fetchUserData(completion: @Sendable @escaping (Result<String, Error>)
-> Void) {
    DispatchQueue.main.asyncAfter(deadline: .now() + 1) {
        completion(.success("User Data fetched!"))
    }
}
```

This method works well in traditional callback-based systems, but it doesn't integrate naturally with async/await. This is where the withCheckedContinuation and withUnsafeContinuation functions can help us. These functions enable us to pause the execution of an async function and resume it later, when a callback completes.

Here's how you could convert the above function to an async version:

```
func fetchUserDataAsync() async throws -> String {
    try await withCheckedThrowingContinuation { continuation in
        fetchUserData { result in
            switch result {
            case .success(let user):
```

```
                    continuation.resume(returning: user)
            case .failure(let error):
                    continuation.resume(throwing: error)
            }
        }
    }
}
```

This `fetchUserDataAsync()` function shows how we can wrap legacy, callback-based API calls to use the modern Swift `async`/`await` interface. Instead of relying on callbacks, it uses `withCheckedThrowingContinuation`, which is Swift's way of letting us "pause" an async function and wait for some legacy asynchronous code to finish.

Inside the continuation block, we call the original `fetchUserData()` function, which still uses a completion handler and returns a result. Once that result comes back, we check whether it was successful or not. If it was successful, we call `resume(returning:)` to continue the async function with the fetched data. If it was not successful, we resume by throwing the error. This function looks and behaves just like any other async function, where we can await it.

This allows you to write cleaner, linear code:

```
do {
    let user = try await fetchUserDataAsync()
    print("User loaded: \(user)")
} catch {
    print("Failed to load user: \(error)")
}
```

There are two types of continuations in Swift:

- `withCheckedContinuation`: Should be used when the task never throws and always returns a valid value
- `withCheckedThrowingContinuation`: Should be used when the task can either return a value or throw an error

We would use the appropriate continuation depending on the original API's design. For example, if we had a function that returned an optional without any error handling, we could write the wrapper like this:

```
func loadImageAsync() async -> UIImage? {
    await withCheckedContinuation { continuation in
        loadImage { image in
            continuation.resume(returning: image)
        }
    }
}
```

Bringing completion handlers into Swift's structured concurrency model is a great way to modernize our code step by step. With continuations, we get cleaner, more readable code and better error handling, without giving up compatibility with older APIs.

Now that we have looked at asynchronous operations and have a good understanding of them, let's take a look at Swift 6's new Strict Concurrency Checking feature.

Strict concurrency checking

One of the significant new features of Swift 6 is the introduction of Strict Concurrency Checking. This feature builds upon the foundation of asynchronous operations concepts such as async/await, tasks, and actors. Strict concurrency checking takes these concepts a step further by enforcing more rigorous compile-time and runtime checks. The main goal of strict concurrency checking is to prevent data race conditions, a common source of bugs in concurrent programming. By implementing stricter checks, Swift 6 aims to catch these issues early in the development process, often at compile time.

There are several key aspects of `Strict Concurrency checking`:

- **Complete data race safety**: Swift 6's compiler will examine the code, looking for potential data race conditions, particularly when multiple tasks are running concurrently. It analyzes not only the individual lines of our code but also how the different parts interact with each other.

- **Sendable enforcement**: The `Sendable` protocol ensures that data types used in concurrent contexts are safe to share across different threads or tasks. A type must conform to `Sendable` either explicitly or implicitly for it to be safely used with asynchronous operations. The compiler evaluates each type's suitability for concurrent use, particularly classes, because they are reference types.

- **Actor isolation**: Actors in Swift control access to their internal state. Swift 6 enforces actor isolation more strictly, ensuring that external code cannot access or modify an actor's state. The compiler checks that all interactions with an actor's state comply with concurrency rules, preventing unauthorized access or modifications.

- **Global variable safety**: Strict Concurrency checking introduces stricter rules for global variables, recognizing that these variables need proper management, especially when multiple threads or tasks might be interacting with them.

> We have several options for managing global variables. We can convert a global variable into a constant, ensuring the value remains unchanged. We might also associate it with a specific actor, effectively isolating its access to a particular context. If we are confident that a variable is safe for concurrent use, we could mark it accordingly, with `nonisolated(unsafe)`, but we must be certain that multiple threads or tasks are not accessing the variable simultaneously.

Strict Concurrency checking is an opt-in feature, and by default, when a new project is started with Swift 6, the setting is for minimal checks. There are several ways to enable complete checks:

- **Swift compiler**: If we are using the Swift compiler, we can pass the `-strict-concurrency=complete` flag like this:

```
swift -strict-concurrency=complete main.swift
```

- **SwiftPM projects**: For projects that use the Swift Package Manager, we would need to add the following to our target's settings in the package manifest file:

```
target(
    name: "MyTarget",
    swiftSettings: [
        .enableUpcomingFeature("StrictConcurrency")
    ]
)
```

- **Xcode projects**: When we are using Xcode, we will need to set the **Strict Concurrency Checking** build setting for our project. To do this, in the **Build Settings** of the project, search for "strict concurrency" and change the setting to **Complete**:

Figure 14.1: Setting strict concurrency checking to Complete

Although Strict Concurrency checking is opt-in by default, enabling it is straightforward across various Swift development environments. Adopting Strict Concurrency checking can greatly improve the stability of your applications by catching concurrency issues early in the development process.

Note that enabling strict concurrency checking could potentially add numerous warnings and errors to your code that will need to be checked.

Default actor isolation

With the introduction of SE-0466 in Swift 6.2, we can now opt into running our code on a single actor by default, specifically the MainActor. This effectively reintroduces a single-threaded programming model, where most code executes on the main actor unless directed otherwise. By doing so, developers can avoid dealing with Swift's concurrency until it becomes necessary.

To enable this behavior, simply set the -default-isolation compile flag to MainActor. The following shows how we can do this with the Swift compiler, SwiftPM projects, and within Xcode:

- **Swift compiler**: If we are using the Swift compiler, we can pass the -default-isolation flag like this:

```
swift -default-isolation=MainActor
```

- **SwiftPM projects**: For projects that use the Swift Package Manager, we would need to add the following to our target's settings in the package manifest file:

```
target(
    name: "MyTarget",
    swiftSettings: [
        .defaultIsolation(MainActor.self)
    ]
)
```

- **Xcode projects**: When we are using Xcode, we will need to set the default actor isolation compiler setting for our project. To do this, in the **Build Settings** of the project, search for "isolation" and change the Default Actor Isolation setting to MainActor:

Figure 14.2: Default Actor Isolation

The Default Actor Isolation feature is an exciting feature, and while it may seem like it is going against all of the Swift concurrency work that has been done since Swift 5.7, it actually solves a very significant issue, which is that Swift concurrency is not needed for many applications.

Summary

We began this chapter by looking at what data race conditions are and how Swift 6 works to prevent them.

Swift's async and await keywords offer a powerful and intuitive way of handling asynchronous tasks, such as network requests and file I/O, ensuring a smooth and responsive user experience. Using the async keyword enables functions to pause and wait while asynchronous operations run, while the await keyword suspends the function until the awaited task completes. This keeps the main thread free, maintaining a responsive user interface.

Tasks and task groups offer additional control and coordination for asynchronous operations. A task is a unit of work that runs asynchronously, enabling concurrent execution without blocking the main thread. Task groups manage related tasks together, enabling multiple operations to run concurrently and gathering their results collectively. By using `async let`, tasks can start simultaneously and be awaited together, ensuring the application continues only when all tasks are completed. This structured concurrency ensures predictable and manageable execution.

Actors are a key feature in Swift's concurrency model. They are designed to safely manage state in concurrent environments and ensure only one thread accesses an actor's mutable state at a time, preventing data races and simplifying concurrent code development. Combined with task groups and the `Sendable` protocol, actors provide a robust framework for writing safe, efficient, and reliable concurrent applications. These features together enable us to build responsive and well-coordinated Swift applications.

We concluded the chapter by looking at the new strict concurrency checking feature introduced in Swift 6 and learned how to enable it in our projects.

In the next chapter, we will look at memory management.

Unlock this book's exclusive benefits now

Scan this QR code or go to `packtpub.com/unlock`, then search this book by name.

Note: Keep your purchase invoice ready before you start.

15

Memory Management

Proper memory management is crucial in application development. It ensures that resources are used efficiently and keeps applications running smoothly. Poor memory management can lead to problems such as memory leaks, slow performance, and even application crashes.

In Swift, memory is managed automatically by a feature called Automatic Reference Counting, or ARC for short. Unlike manual memory management systems, where developers have to allocate and deallocate memory manually, ARC simplifies this process by automatically keeping track of and managing the memory used by class instances. This automation reduces the likelihood of memory-related bugs, enabling us to focus more on the application logic than memory management details.

Despite ARC's automation, we still need to be aware of how ARC works to avoid common pitfalls such as strong reference cycles, which can prevent instances from being deallocated automatically. Understanding the nuances of ARC, including strong, weak, and unowned references, is essential for effective memory management.

In this chapter, we will learn:

- How ARC works
- What a strong reference cycle is
- How to use weak and unowned references to prevent strong reference cycles

Introducing ARC

As we learned in *Chapter 5, Value and Reference Types*, structures and enumerations are value types, whereas classes are reference types. This distinction impacts how they are managed in memory with Swift.

When an instance of a value type is passed within an application, as with a parameter of a method, a copy of the instance is created in memory. This new instance is only valid within the scope where it was created. Once the instance goes out of scope, it is automatically destroyed, and the memory is released. This automatic cleanup makes memory management of value types straightforward and hassle-free.

Classes, on the other hand, operate differently as they are reference types. Memory for an instance of a class is allocated at the time of its creation. When an instance of a class is passed within an application, a reference to the memory location where the instance is stored is passed. Since a class instance may be referenced in multiple scopes, it cannot be automatically destroyed when it goes out of scope. This means the memory is not released; therefore, Swift needs a mechanism to track and release the memory used by class instances when they are no longer needed.

To track and release memory for reference types, Swift implemented ARC. ARC operates mostly automatically, enabling developers to concentrate on other aspects of application development without the need to manage memory manually. However, there are certain situations in which ARC requires additional information to manage memory properly, especially to prevent issues such as strong reference cycles. Before looking into those scenarios, let's start by looking at how ARC works in detail.

How ARC works

Whenever new instances of a class are created, ARC allocates the necessary memory to store those instances. This allocation ensures that there is enough memory to hold the instance's information and locks the memory to prevent it from being overwritten by other processes.

When a class instance is no longer needed, ARC releases the allocated memory so it can be used for other purposes. This process prevents memory from being tied up unnecessarily, which is essential for maintaining optimal performance. If memory is reserved for instances that are no longer needed, it results in a memory leak, which can degrade application performance and stability over time.

On the other hand, if memory is prematurely released for an instance of a class that is still in use, retrieving the class information from memory becomes impossible. Attempting to access an instance after it has been released may lead to application crashes or data corruption. To prevent this, ARC maintains a reference count for each instance, tracking how many active references there are to it. As long as the reference count is greater than zero, the instance remains in memory. Once the reference count drops to zero, indicating that there are no longer any references to the instance remaining, the memory is marked for release.

In playground environments, the previous examples in this book would run properly; however, the following examples in this chapter may not. When running sample code in a playground, ARC does not release objects that we create by design, allowing us to observe the application's behavior and the state of objects at each step. Therefore, to see ARC in action, it's best to run the examples in this chapter in an iOS or macOS project, either through Xcode or from the command line.

Let's look at an example of how ARC works by creating a simple `MyClass` class with the following code:

```
class MyClass {
    var name = ""
    init(name: String) {
        self.name = name
        print("Initializing class with name \(self.name)")
    }
    deinit {
        print("Releasing class with name \(self.name)")
    }
}
```

In this class, we have a `name` property that is set within the initializer. Inside the initializer, a message is printed to the console to indicate that the instance is being created. This class also has a deinitializer, which is called just before an instance of the class is destroyed and removed from memory. The deinitializer prints a message to the console, informing us that the instance is about to be deallocated.

Now, let's look at some code that shows us how ARC creates and destroys instances of a class:

```
var class1ref1: MyClass? = MyClass(name: "One")
var class2ref1: MyClass? = MyClass(name: "Two")
var class2ref2: MyClass? = class2ref1
print("Setting class1ref1 to nil")
class1ref1 = nil
print("Setting class2ref1 to nil")
class2ref1 = nil
print("Setting class2ref2 to nil")
class2ref2 = nil
```

In the example, we begin by creating two instances of the `MyClass` class, named `class1ref1` (which stands for class 1, reference 1) and `class2ref1` (which stands for class 2, reference 1). We then create a second reference to `class2ref1`, named `class2ref2`.

Now, to see how ARC works, we need to set the references to `nil`. We begin this by setting `class1ref1` to `nil`. Since there is only one reference to `class1ref1`, the deinitializer is called. In our case, once the deinitializer completes its task, it prints a message to the console, letting us know that the instance of the class has been destroyed and the memory has been released.

We then set `class2ref1` to `nil`, but there is a second reference to this class (`class2ref2`) that prevents ARC from destroying the instance so that the deinitializer is not called.

Finally, we set `class2ref2` to `nil`, which allows ARC to destroy this instance of the `MyClass` class.

When this code is run, we see the following output, which illustrates how ARC works:

```
Initializing class with name One
Initializing class with name Two
Setting class1ref1 to nil
Releasing class with name One
Setting class2ref1 to nil
Setting class2ref2 to nil
Releasing class with name Two
```

From the example, it seems that ARC handles memory management very well. However, it is possible to write code that will prevent ARC from working properly. Let's take a look at this.

Strong reference cycles

A **strong reference cycle**, or **strong retain cycle**, occurs when two or more objects hold strong references to each other, preventing them from being deallocated. This happens because each object's reference count never reaches zero; their mutual references keep their reference counts above zero, preventing the deallocation of the instances. To avoid this, Swift provides mechanisms such as weak and unowned references, which break the strong reference cycle by not increasing the reference count of the objects they refer to. Managing these references properly ensures proper memory usage and prevents memory leaks in applications.

Now, let's look at an example to see what strong reference cycles are. In this example, we start by creating two classes, named `MyClass1_Strong` and `MyClass2_Strong`, with the following code:

```
class MyClass1_Strong {
    var name = ""
    var class2: MyClass2_Strong?
    init(name: String) {
        self.name = name
        print("Initializing class1_Strong with name \(self.name)")
    }
    deinit {
        print("Releasing class1_Strong with name \(self.name)")
    }
}

class MyClass2_Strong {
    var name = ""
    var class1: MyClass1_Strong?
    init(name: String) {
        self.name = name
        print("Initializing class2_Strong with name \(self.name)")
    }
    deinit {
        print("Releasing class2_Strong with name \(self.name)")
    }
}
```

As we can see from the code, MyClass1_Strong contains an instance of MyClass2_Strong; therefore, the instance of MyClass2_Strong cannot be released until MyClass1_Strong is destroyed. We can also see that MyClass2_Strong contains an instance of MyClass1_Strong; therefore, the instance of MyClass1_Strong cannot be released until MyClass2_Strong is destroyed. This creates a cycle of dependency in which neither instance can be destroyed until the other is.

Let's look at this by running the following code:

```
var class1: MyClass1_Strong? = MyClass1_Strong(name: "Class1_Strong")
var class2: MyClass2_Strong? = MyClass2_Strong(name: "Class2_Strong")
class1?.class2 = class2
class2?.class1 = class1
print("Setting classes to nil")
class2 = nil
class1 = nil
```

In this example, we create instances of the MyClass1_Strong and MyClass2_Strong classes. We then set the class2 property of the class1 instance to the MyClass2_Strong instance. We also set the class1 property of the class2 instance to the MyClass1_Strong instance, meaning that the MyClass1_Strong instance cannot be destroyed until the MyClass2_Strong instance is destroyed. This means that the reference counters for each instance will never reach zero; therefore, ARC cannot destroy the instances, producing the following output:

```
Initializing class1_Strong with name Class1_Strong
Initializing class2_Strong with name Class2_Strong
Setting classes to nil
```

This inability to destroy instances may lead to memory leaks, where an application continues to use memory and does not properly release it. This can cause an application's performance to degrade and possibly lead to application crashes.

To resolve a strong reference cycle, we need to prevent one of the classes from keeping a strong hold on the instance of the other class, thereby allowing ARC to destroy them both. Swift provides two ways of doing this: letting us define the properties as either a weak or an unowned reference.

The difference between a weak reference and an unowned reference is that the instance that a weak reference refers to can be nil, whereas the instance that an unowned reference refers to cannot be nil. This means that when we use a weak reference, the property must be an optional property.

Let's see how we can use unowned and weak references to resolve a strong reference cycle, starting by looking at the unowned reference.

Unowned references

Unowned references are references that do not increase the reference count of the object they refer to. They are used to prevent strong reference cycles in situations where the referenced object has a shorter lifecycle than the referencing object because, unlike weak references, unowned references assume the referenced object will always be in memory when accessed and, therefore, should not be nil. This makes them suitable for scenarios where the referenced object is guaranteed to outlive the reference, ensuring safe memory management.

An unowned reference should only be used when we are certain that the referenced instance will never be deallocated. If we try to access the value of an unowned reference after the instance has been deallocated, we will get a runtime error, and our application will likely crash.

Let's create two more classes, MyClass1_Unowned and MyClass2_Unowned:

```
class MyClass1_Unowned {
    var name = ""
    unowned let class2: MyClass2_Unowned
    init(name: String, class2: MyClass2_Unowned) {
        self.name = name
        self.class2 = class2
        print("Initializing class1_Unowned with name \(self.name)")
    }
    deinit {
        print("Releasing class1_Unowned with name \(self.name)")
    }
}

class MyClass2_Unowned {
    var name = ""
    var class1: MyClass1_Unowned?
    init(name: String) {
        self.name = name
        print("Initializing class2_Unowned with name \(self.name)")
    }
    deinit {
        print("Releasing class2_Unowned with name \(self.name)")
    }
}
```

The MyClass1_Unowned class looks pretty similar to the MyClass1_Strong and MyClass2_Strong classes in the preceding example. The difference here is that with the MyClass1_Unowned class, we set the class2 property to unowned, which means it cannot be nil, and it does not keep a strong reference to the instance that it is referring to. Since the class2 property cannot be nil, we also need to set it when the class is initialized.

Let's see how we can initialize and deinitialize the instances of these classes with the following code:

```
let class2 = MyClass2_Unowned(name: "Class2_Unowned")
let class1: MyClass1_Unowned? = MyClass1_Unowned(
    name: "Cass1_Unowned",class2: class2)

class2.class1 = class1
print("Classes going out of scope")
```

Here, we create an instance of the `MyClass2_Unowned` class and then use that instance to create an instance of the `MyClass1_Unowned` class. We then set the `class1` property of the `MyClass2` instance to the `MyClass1_Unowned` instance we just created.

This creates a reference cycle of dependency between the two classes; however, this time, the `MyClass1_Unowned` instance does not keep a strong hold on the `MyClass2_Unowned` instance, enabling ARC to release both instances when they are no longer needed.

When this code is run, we see the following output, showing that both the `class1` and `class2` instances are released and the memory is properly freed:

```
Initializing class2_Unowned with name Class2_Unowned
Initializing class1_Unowned with name class1_Unowned
Classes going out of scope
Releasing class2_Unowned with name Class2_Unowned
Releasing class1_Unowned with name class1_Unowned
```

As we can see, both instances are properly released.

Now, let's look at how a weak reference can be used to prevent a strong reference cycle.

Weak references

Like unowned references, a weak reference does not increase the reference count of the object it refers to, enabling the object to be deallocated even if there are weak references to it. Weak references must always be optional because the referenced object can be deallocated, leaving the reference with a value of nil.

Once again, to see how this works, we begin by creating two new classes:

```swift
class MyClass1_Weak {
    var name = ""
    var class2: MyClass2_Weak?
    init(name: String) {
        self.name = name
        print("Initializing class1_Weak with name \(self.name)")
    }
    deinit {
        print("Releasing class1_Weak with name \(self.name)")
    }
}

class MyClass2_Weak {
    var name = ""
    weak var class1: MyClass1_Weak?
    init(name: String) {
        self.name = name
        print("Initializing class2_Weak with name \(self.name)")
    }
    deinit {
        print("Releasing class2_Weak with name \(self.name)")
    }
}
```

The MyClass1_Weak and MyClass2_Weak classes look very similar to the previous classes we created, which showed how a strong reference cycle works. The difference is that we define the class1 property in the MyClass2_Weak class as a weak reference.

Now, let's see how we can initialize and deinitialize instances of these classes with the following code:

```swift
let class1: MyClass1_Weak? = MyClass1_Weak(name: "Class1_Weak")
let class2: MyClass2_Weak? = MyClass2_Weak(name: "Class2_Weak")
class1?.class2 = class2
class2?.class1 = class1
print("Classes going out of scope")
```

Here, we create instances of the MyClass1_Weak and MyClass2_Weak classes and then set the properties of those classes to point to the instance of the other class. Once again, this creates a cycle of dependency, but since we set the class1 property of the MyClass2_Weak class to weak, it does not create a strong reference, allowing both instances to be released.

If we run the code, we will see the following output, showing that both the class1_Weak and class2_Weak instances are released and the memory is freed:

```
Initializing class1_Weak with name Class1_Weak
Initializing class2_Weak with name Class2_Weak
Classes going out of scope
Releasing class1_Weak with name Class1_Weak
Releasing class2_Weak with name Class2_Weak
```

> A retain cycle for a closure is exactly the same as a strong reference cycle; a closure is actually a strong reference by default. We would use weak and unowned references to prevent this, exactly as explained in this chapter, simply by changing the variable that holds an instance of a class to hold an instance of a closure.

Starting with Swift 6.2, we also have the ability to create a weak reference to a constant using weak let when declaring a constant property. One thing that is important to note is that using weak let means that the property can't be changed after creation; however, it can still be destroyed.

It is recommended that you avoid creating circular dependencies, as shown in this section, but there are times when they are needed, as in the case of a double-ended linked list. For those times, remember that ARC needs some help to properly manage the memory.

Closures are incredibly powerful, but they can also introduce memory issues similar to the strong reference cycle that we saw in this section. Let's take a look at how we can avoid these memory leaks.

Retain cycles and closures

As we saw in *Chapter 2, Closures and Result Builders*, closures can capture and retain references to any variable or constant from the context in which they were created. This can easily introduce memory issues if not handled correctly in the same way as strong reference cycles, which we just saw.

By default, closures capture string references to any variables or objects they reference from their surrounding context. This includes instances of classes. Let's look at an example of this:

```swift
class Logger {
    var message: String
    var logAction: (() -> Void)?

    init(message: String) {
        self.message = message
    }

    func setupLogging() {
        logAction = {
            print("Log message: \(self.message)")
        }
    }

    deinit {
        print("Logger deinitialized")
    }
}

func runLogger() {
    let logger = Logger(message: "Process started")
    logger.setupLogging()
}
```

In this example, the Logger instance sets a closure to the logAction property, and that closure captures self so it can access the message property. However, since logAction is a property of the Logger instance, we now have a retain cycle where the Logger instance retains the closure, and the closure retains the Logger instance. As a result, the instance can never be deallocated.

To prevent this kind of retain cycle, Swift lets us use capture lists, which allow us to specify how values should be captured inside a closure. For example:

```swift
func setupLogging() {
    logAction = { [weak self] in
        print("Log message: \(self?.message ?? "No message")")
    }
}
```

By capturing self weakly, the closure does not increase the self reference count. This breaks the retain cycle. If self has already been deallocated by the time the closure runs, the weak reference will be nil. We can use both weak and unowned references with the same rules that we saw earlier. The release of Swift 6.2 comes with a new array type called InlineArray. Let's take a look at it.

InlineArray

Introduced in Swift 6.2, InlineArray is a fixed-size array type optimized for performance and memory usage. Unlike Swift's array, which allocates its storage on the heap, InlineArray stores its elements directly within the enclosing type's memory. This inline storage model offers advantages in speed, efficiency, and safety, particularly for small, fixed-size collections.

One of the key benefits of InlineArray is its use of stack memory most of the time instead of the heap, which the standard Array type uses. When an inline array is used as a class property member, it will be allocated inline on the heap with the rest of the properties. An inline array will never use heap memory for its storage alone. Stack memory is significantly faster to allocate and deallocate because it relies on simple pointer arithmetic. When a function is called, memory for its local variables (including any InlineArrays) is reserved by adjusting the stack pointer. Once the function exits, the memory is automatically reclaimed; no manual deallocation or memory management is required. This makes stack memory not only faster but also safer, as it avoids common pitfalls such as memory leaks, which are often associated with heap usage.

Heap memory, by contrast, is used for dynamically sized structures such as Swift's Array. It offers flexibility but comes at a cost: allocation of memory is slower, requires more bookkeeping, and memory must be managed through ARC or other means. Heap allocations are also more prone to fragmentation and less cache-friendly because of the fragmentation. For small collections with a known, fixed size, this level of overhead is often unnecessary, making InlineArray a great choice.

Here's a simple example of InlineArray in use:

```
struct Vehicle {
    var speedSamples: InlineArray<4, Int> = [50, 55, 45, 30]
}

let vehicle = Vehicle()
print("Third speed sample: \(vehicle.speedSamples[2])")
```

In this case, speedSamples is a four-element integer array stored directly within the Vehicle struct without the need for heap allocation. This results in faster access and more predictable memory behavior, which can be a significant performance advantage in environments where we are trying to squeeze as much performance from our code as possible.

That said, InlineArray is not intended to replace Array. It has limitations: the size is fixed and cannot be changed after initialization, and it currently doesn't conform to the Sequence or Collection protocols, so common APIs such as map and for...in are unavailable. But when performance and memory predictability matter, and the size of your collection is known, InlineArray could be a great alternative to the standard array.

Summary

To handle the complexities of memory management for reference types such as classes, Swift employs ARC. ARC tracks the number of references to each instance, automatically releasing the allocated memory when an instance is no longer needed. This prevents memory leaks and keeps our application performance at its best. However, we need to be aware of strong reference cycles, where objects hold strong references to each other, preventing ARC from deallocating them and causing memory leaks.

Swift provides mechanisms such as weak and unowned references to break strong reference cycles. Weak references enable the referenced object to be deallocated, while unowned references assume the object will always be in memory. By properly using these references, we can ensure that our applications are efficiently managing memory and do not have memory leaks.

In the next chapter, we will look at advanced and custom operators in Swift.

Unlock this book's exclusive benefits now

Scan this QR code or go to `packtpub.com/unlock`, then search this book by name.

Note: Keep your purchase invoice ready before you start.

.

16

Advanced and Custom Operators

As software developers, we should be familiar with basic operators such as arithmetic and assignment operators. However, many modern programming languages, including Swift, offer a range of advanced operators, such as bitwise and overflow operators. While these advanced operators may not be as commonly used as their basic counterparts, they are very powerful when used correctly and are particularly useful when working with low-level C-based libraries. These advanced operators also provide essential functionality for tasks requiring precise control and manipulation of data. Understanding and mastering these advanced operators can equip us with the expertise needed to navigate more diverse programming challenges and elevate the quality and effectiveness of our code.

In addition to the built-in basic and advanced operators, we also have the ability to create custom operators. Custom operators enable us to define our own symbols and behavior for operations that are not provided by the standard operators. This allows us to introduce new operators that fit our specific use cases.

In this chapter, you will learn:

- How to use bitwise and overflow operators
- How to write operator methods
- How to create your own custom operator

In this chapter, we'll explore some advanced operators, starting with bitwise operators. But before we dive into these, let's first understand what bits and bytes are.

Bits and bytes

A computer operates using binary digits, commonly known as bits. These bits hold only two possible values: 0 or 1, representing the states of on or off in electrical circuit terms. While bits are tiny and don't have much use on their own except for indicating true/false (Boolean) flags, they become useful when we group them together into sets of 4, 8, 16, 32, or 64 to create data that computers can understand.

In computer terminology, a byte comprises 8 bits. An example of a byte with the value of 42 is depicted as follows, with the least significant bit positioned to the right and the most significant bit to the left:

Number 42	0	0	1	0	1	0	1	0
Bit Values	128	64	32	16	8	4	2	1

Figure 16.1: The number 42 represented in bits

In *Figure 16.1*, the top row illustrates the state of each bit, whether it's off (0) or on (1), within an 8-bit byte. The second row depicts the numerical value assigned to each bit within the byte. In this instance, the bits associated with the values 32, 8, and 2 are set, giving this byte the value of 42. We know this because adding the values of the bits that are set (32, 8, and 2) equals 42.

By default, Swift employs 64-bit numbers; for instance, the standard Int type encompasses 64 individual bits. However, in this chapter, we'll primarily utilize the UInt8 type, which is an unsigned integer containing only 8 bits, or 1 byte, to make it easier to illustrate the examples. It is important to remember that the 64-bit types store bits in the exact same way as a byte; they just contain more bits.

A computer system may store bytes differently in memory, where either the most significant byte or the least significant byte may be located in the lowest memory location. Let's look a little deeper into this concept,

In the preceding example, bits are arranged with the least significant bit on the right and the most significant on the left, as commonly depicted in diagrams. However, in real-life computer systems, bits may be stored differently in memory, where either the most significant bit or the least significant bit is located in the lowest memory address. Let's look deeper into this concept.

Endianness

In computer terminology, endianness refers to the arrangement of bytes in memory. Endianness is typically classified as either big-endian or little-endian. In a little-endian architecture, the least significant byte is stored at the lowest memory address, whereas in a big-endian architecture, the most significant byte occupies the lowest memory address.

When utilizing the Swift standard library or solely working within the Swift language, concerns about bit storage are generally unnecessary. However, when interfacing with low-level C libraries across different architectures, understanding how data is stored becomes essential, particularly when dealing with memory pointers.

For instances where endianness matters, Swift offers built-in instance properties for integers, named littleEndian and bigEndian, to assist in handling endianness-related issues. The following example shows how to use these properties:

```swift
let en = 42
print("Little-endian representation", en.littleEndian)
print("Big-endian representation", en.bigEndian)
```

The en.littleEndian property returns the little-endian encoding of the number 42, while the en.bigEndian property returns the big-endian encoding of the number 42. If we run this code on a device that has a little-endian architecture, we would see the following output:

```
Little-endian representation 42
Big-endian representation 3026418949592973312
```

Based on this output, we can see that the little-endian representation retains the original value that was assigned to the constant. This indicates that the platform we are currently using follows a little-endian architecture.

Both Intel processors and Apple's A and M series processors utilize little-endian architecture for their endianness; therefore, in this chapter, we will assume that everything is little-endian.

Let's next look at what bitwise operators are and how we can use them.

Bitwise operators

Bitwise operators enable us to manipulate the individual bits within a value. A key advantage of these operators is that they are directly supported by the processor; therefore, they can be significantly faster than basic arithmetic operations such as multiplication and division. Later in this chapter, we'll explore how to perform basic multiplication and division using bitwise shift operators.

Before looking into the capabilities of bitwise operators, it's essential to have the means to visualize the binary representation of our variables so that we can see what the operators are doing. Let's examine a couple of methods to accomplish this.

Printing binary numbers

Apple provides us with a generic initializer for the String type that will provide us with the string representation of a given value. This initializer, named `init(_:radix:uppercase:)`, comes with default settings in which uppercase is false and radix is 10. The radix parameter denotes the base of the number to display and has a default value of 10, signifying base 10. To show the binary representation, we must set the radix parameter to 2. By utilizing this initializer, we can display the binary representation of a value in the following manner:

```
let en = 42
print(String(en, radix:2))
print(String(53, radix:2))
```

> Radix is another term for base; therefore, when we set the radix to 2, we are setting the number to base-2 or binary.

If we run this code, we will see the following output:

```
101010
110101
```

In this context, 101010 represents the binary depiction of the number 42, while 110101 represents the binary depiction of the number 53. While this approach works effectively, it does not display leading zeros. For instance, when comparing the binary representations of 53 and 123456, as demonstrated in the following code:

```
print(String(53, radix:2))
print(String(123456, radix:2))
```

we would see the following output:

```
110101
11110001001000000
```

Comparing such representations can be difficult. To conveniently visualize the binary representation of a number, we could integrate the following extension into our codebase:

```
extension BinaryInteger {
    func binaryFormat(_ nibbles: Int) -> String {
        var number = self
        var binaryString = ""
        var counter = 0
        let totalBits = nibbles * 4
        for _ in (1...totalBits).reversed() {
            binaryString.insert(contentsOf: "\(number & 1)",
                               at:binaryString.startIndex)
            number >>= 1
            counter += 1
            if counter % 4 == 0 && counter < totalBits {
                binaryString.insert(contentsOf: " ",
                                   at: binaryString.startIndex)
            }
        }
        return binaryString
    }
}
```

It is OK if you do not understand how this code works at this time, as bitwise shift operators haven't been explained yet. Once they've been covered later in this chapter, you will be able to understand how this code works.

This extension is designed to accept an integer and furnish the binary representation of the number, complete with the requisite number of nibbles. A nibble constitutes half a byte or 4 bits. Within the resulting string, the code inserts a space between each nibble for improved readability. The following code shows how we could use this extension:

```
print(53.binaryFormat(2))
print(230.binaryFormat(2))
```

With this code, we are displaying the binary representation of the numbers 53 and 230 in two nibbles. The following shows the output of this code:

```
0011 0101
1110 0110
```

Now that we have a very basic idea of what bits, bytes, nibbles, and endianness are, and we are able to display numbers in binary format, let's look at bitwise operators, starting with the bitwise AND operator.

The bitwise AND operator

The bitwise AND operator (&) takes two values and returns a new value where the bits in the new value are set to 1 only if the corresponding bits of both input values are set to 1. The AND operator can be read as this: if a bit from the first value AND the corresponding bit from the second value are both 1, then set the corresponding bit of the resultant value to 1.

Let's see how this works by seeing how we would use a bitwise AND operation on the numbers 42 and 11:

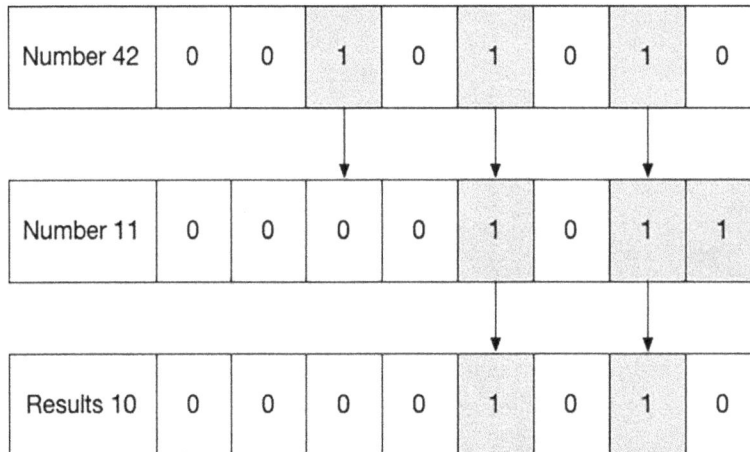

Number 42	0	0	1	0	1	0	1	0

Number 11	0	0	0	0	1	0	1	1

Results 10	0	0	0	0	1	0	1	0

Figure 16.2: The AND operator

As this diagram shows, the second and fourth bits from the right are both set to **1 for the numbers 42 and 11**; therefore, the results of the AND operation have those bits set, resulting in an output value of **10**. Now, let's see how this works in code:

```
let numberOne: Int8 = 42
let numberTwo: Int8 = 11
print("\(numberOne) = \(numberOne.binaryFormat(2))")
print("\(numberTwo) = \(numberTwo.binaryFormat(2))")
let andResults = numberOne & numberTwo
print("\(andResults) = \(andResults.binaryFormat(2))")
```

The previous code assigns the values 42 and 11 to two integers. It then outputs the binary representation of these numbers, formatted in two nibbles, using the `binaryFormat` extension, to the console. Following this, it executes a bitwise AND operation on the integers and prints the binary representation of the resultant values to the console. The following results will be printed to the console:

```
42 = 0010 1010
11 = 0000 1011
10 = 0000 1010
```

As we can see, the result from the code is the same as shown in the diagram, which has a result of 10. Now, let's look at the bitwise OR operator.

The bitwise OR operator

The bitwise OR operator (|) takes two values and returns a new value where the bits of the results are set to 1 only if the corresponding bits of either or both values are set to 1. The OR operation reads as this: if the bit from the first value OR the corresponding bit from the second value is 1, then set the bit in the results to 1.

Let's see how this works by seeing how we would do a bitwise OR operation on the numbers 42 and 11:

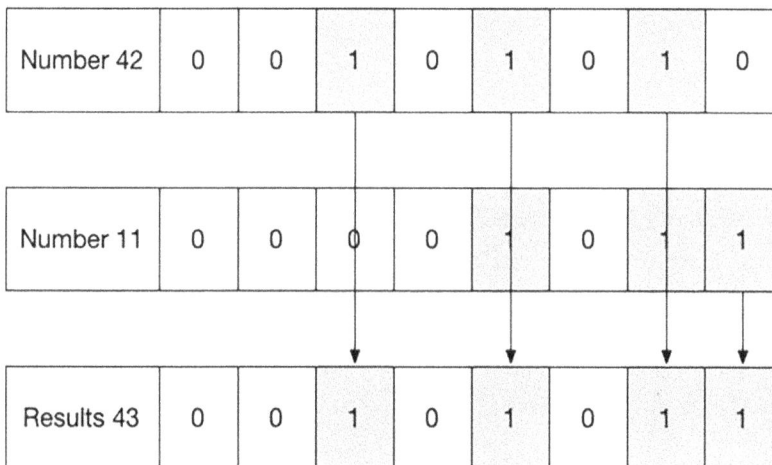

Figure 16.3: The OR operator

As this diagram shows, the first, second, fourth, and sixth bits from the right are set in one or both of the values; therefore, the results of the OR operation have all of those bits set. Now, let's see how this works in code:

```
let numberOne: Int8 = 42
let numberTwo: Int8 = 11
print("\(numberOne) = \(numberOne.binaryFormat(2))")
print("\(numberTwo) = \(numberTwo.binaryFormat(2))")
let orResults = numberOne | numberTwo
print("\(orResults) = \(orResults.binaryFormat(2))")
```

The previous code assigns the values of two integers to 42 and 11. It then outputs the binary representation of these numbers, formatted in two nibbles, using the binaryFormat extension, to the console. Following this, it executes a bitwise OR operation on the integers and prints the binary representation of the resultant values to the console. The following results will be printed to the console:

```
42 = 0010 1010
11 = 0000 1011
43 = 0010 1011
```

As we can see, the result from the code is the same as shown in the diagram, which has a result of 43. Now, let's look at the bitwise XOR operator.

The bitwise XOR operator

The bitwise XOR operator (^) takes two values and returns a new value where the bits of the new value are set to 1 only if the corresponding bits of either but not both input values are set to 1. The XOR operator reads as follows: if the bit from the first value OR the corresponding bit from the second value is 1, but not both, then set the bit of the results to 1.

Let's see how this works by seeing how we would do a bitwise XOR operation on the numbers 42 and 11:

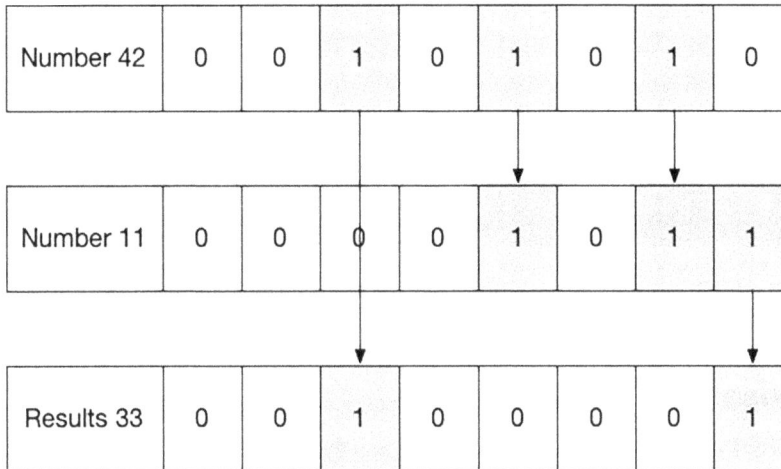

Figure 16.4: The XOR operator

As this diagram shows, the second and fourth bits from the right are set to **1** for both numbers; therefore, in the results, those bits are not set. However, the sixth bit in the number **42** is set to **1** and the first bit in the number **11** is set to **1**; therefore, in the results, those bits are set. Now, let's see how this works in code:

```
let numberOne: Int8 = 42
let numberTwo: Int8 = 11
print("\(numberOne) = \(numberOne.binaryFormat(2))")
print("\(numberTwo) = \(numberTwo.binaryFormat(2))")
let xorResults = numberOne ^ numberTwo
print("\(xorResults) = \(xorResults.binaryFormat(2))")
```

The previous code sets the value of two integers to 42 and 11. It then outputs the binary representation of these numbers, formatted in two nibbles, using the `binaryFormat` extension, to the console. Following this, it executes a bitwise XOR operation on the integers and prints the binary representation of the resultant values to the console. The following results will be printed to the console:

```
42 = 0010 1010
11 = 0000 1011
33 = 0010 0001
```

As we can see, the result from the code is the same as shown in the diagram, which has a result of 33. Now, let's look at the bitwise NOT operator.

The bitwise NOT operator

The bitwise NOT operator (~) is different from the other logical operators because it only takes one value. The bitwise NOT operator will return a value where all the bits are reversed. What this means is that any bit on the input value that is set to 1 will be set to 0 on the resulting value, and any bit that is set to 0 on the input value will be set to 1 on the resulting value. Let's see how this would work given a value of 42:

Number 42	0	0	1	0	1	0	1	0
Results 213	1	1	0	1	0	1	0	1

Figure 16.5: The NOT operator

The diagram illustrates that when we perform the bitwise NOT operation, all the bits in the result's value will be the opposite of what they were in the original value. Let's see what this looks like in code:

```
let numberOne: Int8 = 42
print("\(numberOne) = \(numberOne.binaryFormat(2))")
let notResults = ~numberOne
print("\(notResults) = \(notResults.binaryFormat(2))")
```

The previous code performs the NOT operation on the value of the numberOne variable. The following results will be printed to the console:

```
42 = 0010 1010
-43 = 1101 0101
```

Notice the result of the bitwise NOT operation is a negative number. The reason for this is that an integer is a signed number. With signed numbers, the most significant bit designates whether the number is a positive number or a negative number. With all bits being reversed, a negative number will always turn into a positive number and a positive number will always turn into a negative number. Using an unsigned integer prevents this from happening because the most significant bit represents part of the value itself, rather than being used to indicate the sign.

Now that we have looked at the logical bitwise operators, let's look at the bitwise shifting operators.

Bitwise shift operators

Swift provides two bitwise shift operators: the bitwise left-shift operator (<<) and the bitwise right-shift operator (>>). These operators shift all bits to the left or right by the number of places specified. The shift operators have the effect of multiplying (left-shift operator) or dividing (right-shift operator) by factors of two. By shifting the bits to the left by one, you are doubling the value, and shifting them to the right by one will halve the value. Let's see how these operators work, starting with the left-shift operator:

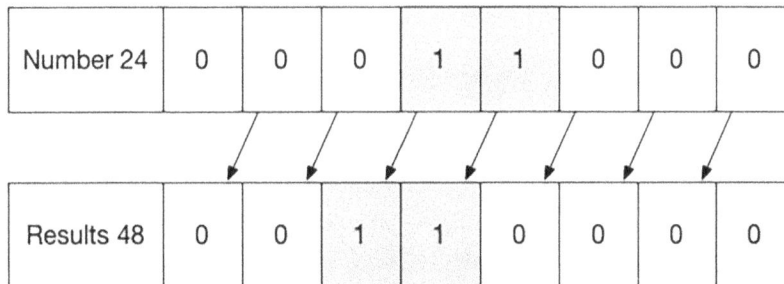

Figure 16.6: The left-shift operator

With the left-shift operator, all bits in the original value are shifted to the left by one, with the most significant bit falling off and not factoring into the final result. The least significant bit in the result will always be set to zero.

Now, let's look at the right-shift operator:

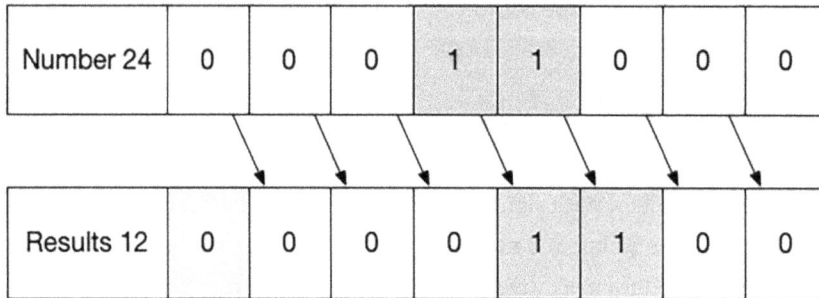

Figure 16.7: The right-shift operator

With the right-shift operator, all bits in the original value are shifted one spot to the right, with the least significant digit falling off. The most significant digit in the result will always be set to zero.

Now, let's see what this looks like in code:

```
let numberOne: UInt8 = 24
let resultsLeft = numberOne << 1
let resultsRight = numberOne >> 1
let resultsLeft3 = numberOne << 3
let resultsRight4 = numberOne >> 4
print("24  \(numberOne.binaryFormat(2))")
print("\(resultsLeft) <<1 \(resultsLeft.binaryFormat(2))")
print("\(resultsRight) >>1 \(resultsRight.binaryFormat(2))")
print("\(resultsLeft3) <<3 \(resultsLeft3.binaryFormat(2))")
print("\(resultsRight4) >>4 \(resultsRight4.binaryFormat(2))")
```

In this code, we start by setting a variable to the number 24. We then use the left-shift operator to shift the bits one spot to the left. The number after the shift operator defines how many spots to shift the numbers. The next line shifts the bits one spot to the right, then the next line shifts the bits three spots to the left, and the next line shifts the bits four spots to the right. The final five lines print out the results to the console. If you run this code, you should see the following results:

```
24 0001 1000
48 <<1 0011 0000
12 >>1 0000 1100
192 <<3 1100 0000
1 >>4 0000 0001
```

Looking at the results, we can see that the bits are shifted to the left or right depending on the shifting operator used. In the last line, we can see that when we shifted to the right four spaces, only one bit was set to 1 rather than two. This is because the bit in the fourth spot from the right in the original number "fell off." If we had shifted to the right five spots, both bits that were set to 1 in the original number would have fallen off, and we would have been left with all zeros.

Now that we understand how the bitwise operators work, let's look at the BinaryInteger extension that we created earlier in this chapter.

BinaryInteger extension

Earlier in the chapter, we created the BinaryInteger extension that contained the binaryFormat() function. The code for this extension is here:

```
extension BinaryInteger {
    func binaryFormat(_ nibbles: Int) -> String {
        var number = self
        var binaryString = ""
        var counter = 0
        let totalBits = nibbles*4
        for _ in (1...totalBits).reversed() {
            binaryString.insert(contentsOf: "\(number & 1)",
                                at:binaryString.startIndex)
            number >>= 1
            counter += 1
            if counter % 4 == 0 && counter < totalBits {
                binaryString.insert(contentsOf: " ",
                                    at: binaryString.startIndex)
            }
        }
        return binaryString
    }
}
```

In this code, we define an extension to the `BinaryInteger` protocol, which adds a method to format an integer as a binary string with a specific number of nibbles (four bit groups) for better readability.

The `binaryFormat()` method takes a single parameter, called `nibbles`, of the integer type, that indicates the number of nibbles to use in the string representation of the number. Inside the method, a `number` variable is initialized: `number`, which is a mutable copy of the integer. We also have `binaryString`, which is an initially empty string that will store the formatted binary representation of the number; `counter`, which tracks the bits processed; and `totalBits`, calculated with `nibbles * 4`, which gives us the total number of bits to format.

A `for` loop is created that loops from `totalBits` down to 1. Inside the loop, the least significant bit of `number` is extracted using the bitwise AND operation and inserted at the start of `binaryString`. The right-shift operator is then used to right-shift the number by one bit, and the counter is incremented by one. For every four bits, a space is inserted at the start of `binaryString` to separate the nibbles. Finally, the method returns the `binaryString`, which is the binary representation of the original number.

Now, let's look at overflow operators.

Overflow operators

Swift, at its core, is designed for safety. One of these safety mechanisms is the inability to insert a number into a variable when the variable type is too small to hold it. As an example, the following code will throw the following error: arithmetic operation '255 + 1' (on type 'UInt8') results in an overflow:

```
let b: UInt8 = UInt8.max + 1
```

The reason an error is thrown is that we are trying to add one to the maximum number that a UInt8 can hold. This error checking can help prevent unexpected and hard-to-trace issues in our applications.

Let's take a second to look at what would happen if Swift did not throw an error when an overflow occurs. In a UInt8 variable, which is an 8-bit unsigned integer, the number 255 is stored like this, where all the bits are set to 1:

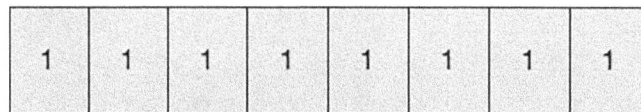

1	1	1	1	1	1	1	1

Figure 16.8: The binary representation of 255

Now, if we add 1 to this number, the new number will be stored like this:

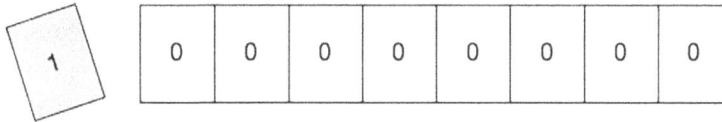

Figure 16.9: Overflow when trying to represent 256

Notice that the 8 bits that represent the UInt8 number are all 0s, while the leading 1 falls off or overflows because we can only store 8 bits. In this case, when we add 1 to the number 255, the number stored in the results would be 0 if we did not have overflow error checking. This could lead to very unexpected behavior in our code that would be hard to track down.

If this is the behavior that we want, because we are not concerned about lost bits, Swift does offer three overflow operators that will allow us to opt into this behavior. These are the overflow addition operator (&+), the overflow subtraction operator (&-), and the overflow multiplication operator (&*). The following code shows how these operators work:

```
let add: UInt8 = UInt8.max &+ 1
let sub: UInt8 = UInt8.min &- 1
let mul: UInt8 = 42 &* 10
print("add: \(add): \(add.binaryFormat(2))")
print("sub: \(sub): \(sub.binaryFormat(2))")
print("mul: \(mul): \(mul.binaryFormat(2))")
```

In this code, we add 1 to the maximum value of the UInt8 type, which is 255, subtract 1 from the UInt8 type minimum value, which is 0, and then multiply 42 by 10, which has a result greater than the 255 maximum value of the UInt8 type. The results that are printed to the console are as follows:

```
add: 0: 0000 0000
sub: 255: 1111 1111
mul: 164: 1010 0100
```

As we can see from the results, when we add 1 to the maximum value of the UInt8 type, the result is 0. When we subtract 1 from the minimum value of the UInt8 type, the result is 255 (the maximum value of the UInt8 type). Finally, when we multiply 42 by 10, which our arithmetic teachers would tell us is 420, we end up with 164 because of the overflow.

Now, let's look at how we can use operator methods to add operators to custom types.

Operator methods

Operator methods enable us to add implementations for standard Swift operators to our custom classes and structures, a process known as operator overloading. This is a very useful feature because it lets us apply common operations to our custom types using well-known operators. Let's explore how to accomplish this, but first, we'll create a custom type, which we will use for our examples, called MyPoint:

```
struct MyPoint {
    var x = 0
    var y = 0
}
```

The MyPoint structure defines a two-dimensional point on a graph.

Now that we have our custom type to work with, let's add three operator methods to this type. These operators include the addition operator (+), the addition assignment operator (+=), and an inverse operator (-). The addition operator and the addition assignment operator are infix operators because there is a left and right operand (value) to the operation, while the inverse operator is a prefix operator because it is used before a single value:

```
extension MyPoint {
    static func + (left: MyPoint, right: MyPoint) -> MyPoint {
        return MyPoint(x: left.x + right.x, y: left.y + right.y)
    }
    static func += (left: inout MyPoint, right: MyPoint) {
        left.x += right.x
        left.y += right.y
    }
    static prefix func -(point: MyPoint) -> MyPoint {
        return MyPoint(x: -point.x, y: -point.y)
    }
}
```

> There are also postfix operators, which are used after a single value.

When incorporating operator methods into our types, we define them as static functions with the operator symbols serving as the method names. For prefix or postfix operators, we also include the "prefix" or "postfix" keyword before declaring the function.

The addition operator is an infix operator; therefore, it takes two input parameters of the MyPoint type. One parameter represents the MyPoint instance on the left side of the operator, while the other represents the instance on the right side.

Similarly, the addition assignment operator, also an infix operator, requires two input parameters of the MyPoint type. However, unlike the addition operator, the resulting value from the addition assignment operation is assigned to the MyPoint instance on the left side of the assignment operator. Therefore, this parameter is specified as an inout parameter to enable the return of results within that instance.

Lastly, the inverse operator is a prefix operator, which is utilized before an instance of the MyPoint type; therefore, it only requires a single parameter of the MyPoint type.

Let's see how these operators work:

```
let firstPoint = MyPoint(x: 1, y: 4)
let secondPoint = MyPoint(x: 5, y: 10)
var combined = firstPoint + secondPoint
combined += firstPoint
let inverse = -combined
```

This code starts by defining two points and then adds them together using the addition operator that was created. The result is stored in the new combined instance of the MyPoint type. The combined instance will contain the values of x as 6 and y as 14.

Next, the addition assignment operator is used to add the values in the firstPoint instance to the values in the combined instance that was created in the first step. The result is stored in the combined instance. After this operation, the combined instance contains the values of x as 7 and y as 18.

Finally, the inverse operator is used with the combined instance to reverse the values and save the new values in the inverse instance of the MyPoint type. The inverse instance contains the values of x as -7 and y as -18.

We are not limited to using only current operators but can also create our own custom operators as well. Let's see how we can do this.

Custom operators

Custom operators offer us the flexibility to introduce and implement our own operators beyond the set provided by Swift. These new operators need to be globally declared using the operator keyword and should be accompanied by either infix, prefix, or postfix keywords for proper definition. Once a custom operator is globally defined, we can integrate it into our types using operator methods, as illustrated in the preceding section.

In Swift, operators can be classified based on their position relative to their operands. These classifications are:

- **infix**: The operator is placed between two operands
- **prefix**: The operator is placed before a single operand
- **postfix**: The operator is placed after a single operand

To look deeper into this concept, let's introduce two new operators that we can integrate into the MyPoint type used in the preceding section:

- The operator •, which will be used for multiplying two points together
- The operator ••, which will be used for squaring a value

The • symbol can be typed by holding down the Option key and pressing the number 8 on a computer running macOS.

The first thing we need to do is to declare the operators globally. We can use the following code to achieve this:

```
infix operator •
prefix operator ••
```

Notice that we specify the type of operator (infix, prefix, or postfix), followed by the keyword operator, and then the symbols designated for the operator. Now, we can utilize these operators in a manner similar to standard operators with our MyPoint type. The following code shows how we would use these operators with the MyPoint type:

```
extension MyPoint {
    static func • (left: MyPoint, right: MyPoint) -> MyPoint {
        return MyPoint(x: left.x * right.x, y: left.y * right.y)
```

```
        }

        static prefix func •• (point: MyPoint) -> MyPoint {
            return MyPoint(x: point.x * point.x, y: point.y * point.y)
        }
    }
```

These new custom operators are added to the MyPoint type exactly as we would add standard operators, using static functions. We are now able to use these operators:

```
    let firstPoint = MyPoint(x: 1, y: 4)
    let secondPoint = MyPoint(x: 5, y: 10)

    let multiplied = firstPoint • secondPoint
    let squared = ••secondPoint
```

In the first line, we use the • operator to multiply two instances of the MyPoint type together. The results are stored in the multiplied instance of the MyPoint type. The multiplied instance will now contain the values of x as 5 and y as 40.

We then use the •• operator to square the value of the secondPoint instance and put the new value in the squared instance. The squared instance will now contain the values of x as 25 and y as 100.

Summary

In this chapter, we learned how we can use the advanced bitwise AND, OR, XOR, and NOT operators to manipulate the bits of values stored in variables. We also looked at how we can use the left- and right-shift operators to shift bits to the left and right.

We examined the role of overflow operators in modifying the default behavior of arithmetic operations, preventing errors in cases where results exceed the maximum or minimum values for a given type.

We also looked at the concept of adding operator methods to types, enabling us to use the standard operators provided by Swift with our custom types. Additionally, we learned how to create our own custom operators, as well.

In the next chapter, we will look at access controls in Swift.

Unlock this book's exclusive benefits now

Scan this QR code or go to packtpub.com/unlock, then search this book by name.

Note: Keep your purchase invoice ready before you start.

17

Access Controls

Access control is a key feature that enables developers to manage the visibility and accessibility of different parts of their code. Access control in Swift is designed to prevent unauthorized access to certain parts of our codebase and ensure that components interact only as intended. This helps us maintain secure and reliable applications and frameworks.

Swift defines five distinct levels of access control: open, public, internal, file-private, and private. These levels determine the visibility of classes, structures, enumerations, methods, and other entities within our Swift code.

By selecting the appropriate access levels, we can organize our code more effectively, reducing the risk of unintended interactions while maintaining a clear and logical structure within our codebase. Understanding Swift's access control mechanisms allows you to write cleaner, safer, and more modular code.

In this chapter, we will learn:

- What access controls are and how they are used
- How we can restrict access using the different access control levels
- What the best practices are when using access controls

Introducing access control

Access control is a vital part of software development that ensures the security and integrity of code by limiting who can access and change certain parts of it. By controlling what can be seen and modified within our code, we can create organized and well-protected applications and frameworks. This makes the code easier to use, maintain, and secure.

Without proper access controls, unauthorized modifications of properties or methods can lead to significant issues. These modifications may create security vulnerabilities, introduce unexpected behavior, or even cause the application to crash. By restricting access, we ensure that each component behaves as intended, maintaining the software's integrity, security, and overall stability. In the end, good access control makes software more reliable and secure.

Access control levels

Access control in Swift is implemented with access control levels. These levels determine where the entities, such as types, properties, methods, and initializers, can be accessed from within our code. There are five access control levels. Let's take a look at them.

Open access

The open access level is the most permissive and is used for class and class members. Entities marked as open can be accessed and subclassed from anywhere, including other modules. This level of access is typically used for framework classes that are intended to be subclassed by client code.

The following code creates a class with a single property and method that are defined with the open access level:

```swift
open class OpenClass {
    open var openProperty: Int

    init(openProperty: Int) {
        self.openProperty = openProperty
    }

    open func openMethod() { }
}
```

The open access level can only be used by classes and overridable class members. For all other entities, you will want to use the public access level.

> ♀ **Quick tip**: Enhance your coding experience with the **AI Code Explainer** and **Quick Copy** features. Open this book in the next-gen Packt Reader. Click the **Copy** button
>
> (1) to quickly copy code into your coding environment, or click the **Explain** button
>
> (2) to get the AI assistant to explain a block of code to you.

```
function calculate(a, b) {                                    Copy      Explain
    return {sum: a + b};                                        1          2
};
```

> The next-gen **Packt Reader** is included for free with the purchase of this book. Scan the QR code OR go to packtpub.com/unlock, then use the search bar to find this book by name. Double-check the edition shown to make sure you get the right one.

Public access

The public access level is the default for top-level entities such as types, global constants, and variables in a module. Public entities can be accessed from anywhere, including other modules that import the defining module.

The following code shows how we would use the public access level, through the creation of a structure with a single property and method that are defined with the public access level:

```
public struct PublicStruct {
    public var publicProperty: String
    public func publicMethod() { }
}
```

Internal access

The internal access level is the default for entities that are not explicitly marked with any other access control level. Internal entities can be accessed within the same module, but not from outside the module.

The following code creates a structure with a single property and method that are defined with the internal access level:

```
struct InternalStruct {
    internal var internalProperty: Double
    var internalProperty2: Int
    func internalMethod() { }
}
```

Notice that we are only using the `internal` access modifier for the first property but not for the second property or the function. That is because the internal access level is the default for entities that are not explicitly marked.

When creating Swift packages, all entities are internal to the package by default. This means that using the internal access control level only enables properties and methods to be accessible within the package itself. In order to make these entities available to your application, you must declare them as public.

File-private access

Entities marked as file-private can only be accessed from within the same source file. This access level is useful for hiding implementation details and providing a clear separation of concerns within a specific file.

The following code creates a structure with a single property and method that are defined with the file-private access level:

```
fileprivate struct FilePrivateStruct {
    var filePrivateProperty: Bool
    func filePrivateMethod() { }
}
```

Private access

The private access level is the most restrictive in Swift. Private entities can only be accessed from within the same type in which they are defined or extensions of that type. This level of access is commonly used to encapsulate and protect the internal state of a type.

The following code creates a structure with a single property and method that are defined with the private access level:

```
struct myStruct {
    private var privateProperty: Int
    private func privateMethod() { }
}
```

If we define a type, such as a structure or class, with the private access level, it will be accessible to code within the same source file it is defined in.

Now let's look at how we can use access controls with enumerations.

Access levels with enumerations

We can apply access controls to enumerations, which will restrict or expose their visibility based on our design requirements. When we set an access level for an enumeration, all of its cases automatically inherit that access level. This ensures that the enumeration and its cases are consistently controlled.

As with other entities, the default access control for enumerations that are not explicitly marked is internal. The following code shows how we would use access controls with enumerations:

```
public enum Direction {
    case north
    case south
    case east
    case west
}
```

In this example, we declare that the access level for the `Direction` enumeration is public.

Raw values or associated values that may be of a custom type, in an enumeration definition, must have an access level at least as high as the enumeration's access level. As an example, we can't use a type that has an access level of private as the raw-value type of an enumeration with an open access level.

Next, let's see how we can enable more fine-grained access control over the getters and setters of our properties.

Access levels for getters and setters

Getters and setters for properties and subscripts will automatically receive the same access level as the property or subscript that they belong to; however, a setter can have a lower access level than its corresponding getter. Let's look at an example of this:

```
struct SampleStruct {
    private(set) var count = 0

    mutating func doSomething() {
        count += 1
    }
}
```

In the preceding example, we assign the private access level to the count variable using the private(set) modifier. Other modifiers that we can use are the fileprivate(set) or the internal(set) modifiers. We could use this structure like this:

```
var test = SampleStruct()
test.doSomething()
test.doSomething()
print(test.count)
```

This code would print out the value 2 to the console because we called the doSomething() method twice. If we attempted to set the count properties as shown in the following example, we would receive an error that the property was not accessible:

```
test.count = 5
```

The reason the error would be thrown is we assigned a private access level to the setter of the property, meaning it is only accessible from the same type in which it is defined or any extensions of that type.

Best practices for access control

Swift's access control system offers flexibility in managing the visibility and encapsulation of your code. However, it is important to adhere to best practices to help ensure your code is easily maintainable, secure, and in alignment with software design principles. In this section, we will cover some recommended best practices for using access controls in Swift.

Use the most restrictive access level by default

Always start with the most restrictive or deny all stance (private), and then selectively open up access only to the extent required. This minimalistic strategy, known as the "principle of least privilege," ensures that we expose only the essential components while keeping the rest of the implementation details concealed. By limiting access to what is absolutely necessary, we reduce the risk of unintended access or modification of our codebase.

This approach promotes better modularity and maintainability of our applications and frameworks, making it easier to understand and manage our code in the future. It helps prevent accidental misuse and keeps the code's internal logic encapsulated, thus adhering to best practices of software design principles.

Encapsulate implementation details

Encapsulation is a fundamental principle of object-oriented and protocol-oriented programming. Access controls play a crucial role in achieving encapsulation by hiding implementation details and only exposing a well-defined public interface. By only exposing a well-defined public interface, we can ensure that our code is easier to maintain, extend, and refactor without breaking client code.

We should use private and file-private access levels to encapsulate implementation details that should not be publicly exposed. Use the internal access level for details that should only be accessed within the module in which they are developed. The public access level should only be used for details that are part of our documented public interface.

Use extensions wisely

When adding functionality to a type with an extension, it is crucial to align the access level of the extension with the proper scope for the added functionality. This practice will ensure that the visibility of the new methods and properties is consistent with their intended use and ensures better-organized and more maintainable code, as each part of the codebase only interacts with the parts it needs to.

Maintain consistency

Establishing and following a consistent access control strategy throughout your codebase is essential to ensure that your code is easy to understand and maintain.

To maintain consistency, we should define and document our access control guidelines and conventions for our projects and teams. We could also use tools such as code linters or static analysis tools to enforce consistent access control practices across our codebase.

Summary

Understanding and properly using access controls is essential in software development to ensure code security and integrity by restricting access to certain parts of the codebase. By using proper access levels with our code, we can prevent unauthorized access and modification to specific parts of our code, ensuring that each part of the application or framework operates as intended.

Swift implements access control using five access levels, which are open, public, internal, file-private, and private. These levels dictate the visibility and accessibility of entities such as types, properties, methods, and initializers. Starting with the most restrictive level (private) and selectively opening up access as needed, following the "principle of least privilege," minimizes exposure of the implementation details of our code. This strategy reduces risks and promotes better modularity and maintainability.

In the next chapter, we will look at object-oriented programming with Swift.

Unlock this book's exclusive benefits now

Scan this QR code or go to packtpub.com/unlock, then search this book by name.

Note: Keep your purchase invoice ready before you start.

18

Swift Testing

Unit tests offer many advantages in software development. They help ensure code reliability by catching bugs early, make refactoring easier by providing a safety net, and improve overall code quality. Unit tests also serve as documentation by showing how the code is intended to be used, and they boost developer confidence, allowing for faster iterations and more robust feature development.

Swift 6 introduces a new testing framework called Swift Testing, designed to transform how developers write and organize tests for Swift code. This open-source package features a modern, expressive API that leverages Swift's latest features, including concurrency and macros, to create more efficient and readable test suites.

Swift Testing is a major advancement in the language's evolution. It simplifies the testing process with tools for describing and organizing tests, delivering detailed failure reports, and supporting workflows such as conditional testing and parameterized tests. With its open development process, Swift Testing also invites community participation, allowing us to shape its future and contribute to its growth as a core component of the Swift ecosystem.

In this chapter, we will learn:

- How to add Swift Testing to our projects
- How to create unit test functions
- How to use the #expect macro
- How to use traits and suites

To use Swift Testing, we will need to ensure our project has a testing target. If your project does not already include a testing target, the next section will explore how to add one.

Getting started with Swift Testing

Swift Testing and the older XCTest can be built within the same test target and coexist within the same test bundle. If our project already contains an XCTest bundle, there's no need to create a new unit-testing bundle to use Swift Testing. We can add new Swift Testing files to the existing target and, over time, convert existing XCTest-based tests to Swift Testing.

There are multiple methods for adding a Swift Testing target to a project. In this section, we will look at how to add a Swift Testing target to both an existing Xcode project and an existing Swift project created with the Swift Package Manager. Let's start by adding a testing target to an existing Xcode project.

Adding Swift Testing to an existing Xcode project

The following steps outline how to add Swift Testing to an existing Xcode project.

1. From the top menu, select **File** | **New** | **Target...**:

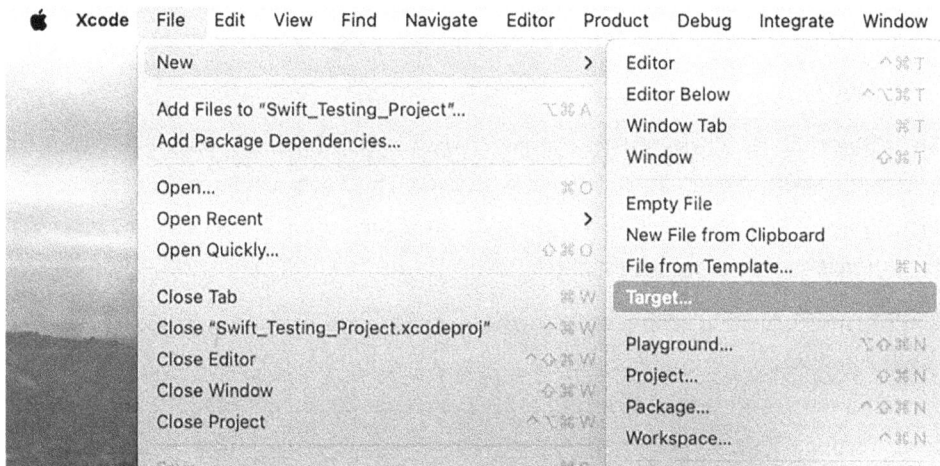

Figure 18.1: Selecting Target...

2. In the template filter box at the top right of the **Target** window, type test and select **Unit Testing Bundle** or **UI Testing Bundle**, according to your needs.

Choose a template for your new target:

Multiplatform   iOS   macOS   watchOS   tvOS   visionOS   DriverKit   Other   test ⊗

Test

UI Testing Bundle   **Unit Testing Bundle**

Cancel   Next

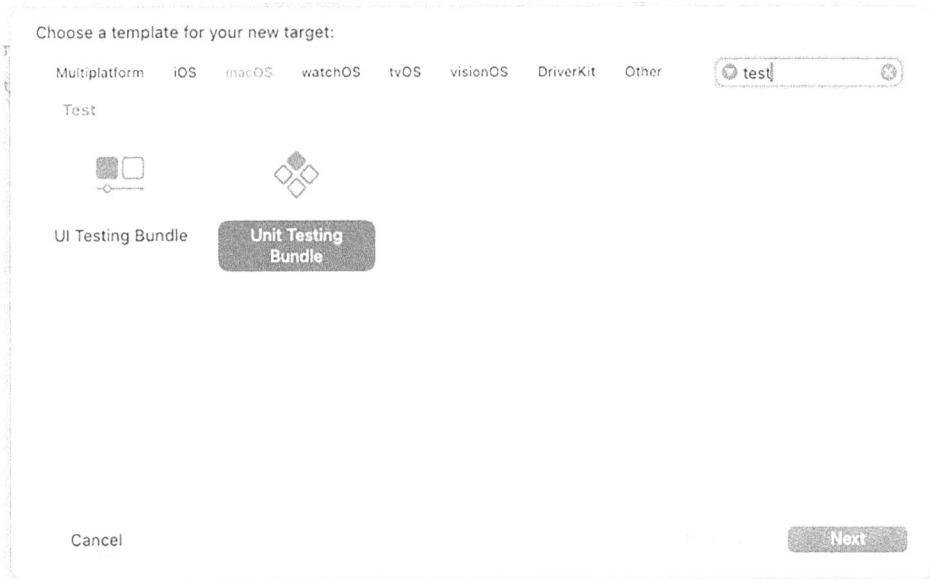

Figure 18.2: Template selection window

3. Fill out the options and click **Finish**.

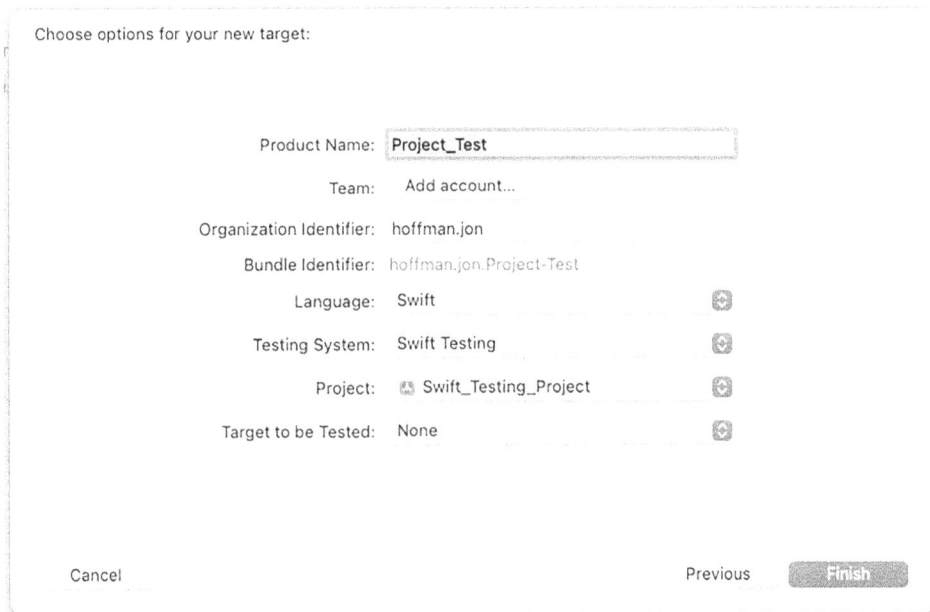

Choose options for your new target:

Product Name: **Project_Test**

Team: Add account...

Organization Identifier: hoffman.jon

Bundle Identifier: hoffman.jon.Project-Test

Language: Swift

Testing System: Swift Testing

Project: Swift_Testing_Project

Target to be Tested: None

Cancel   Previous   Finish

Figure 18.3: New target options window

4.   After selecting **Finish**, a Swift Testing target will be added to your project. This target will allow you to write and run tests within your Xcode environment. The following screenshot shows how the testing target will appear within the Xcode environment:

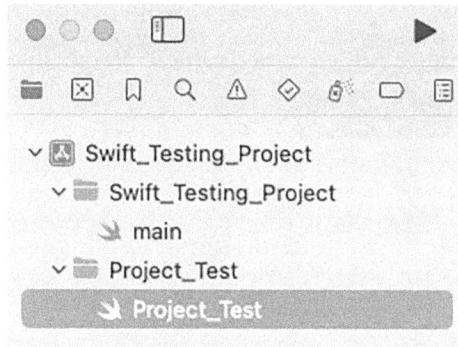

Figure 18.4: Test target in a project

Now that we have seen how to add a Swift Testing target to an Xcode project, let's see how to add one to a project created with the Swift Package Manager from the command line.

Adding Swift Testing to an existing Swift Package Manager project

To add a Swift Testing target to a project created with the Swift Package Manager, we need to modify the `Package.swift` file by including a few additional entries. This involves specifying the test target, linking it to the Swift Testing package, and adding dependencies.

To do this, we begin by adding the Swift Testing package as a dependency. Insert the following code into your `Package.swift` file. If you already have other dependencies, simply add the `.package` line to the existing `dependencies` block as you do not want to create a new one:

```
dependencies: [
  package(url: "https://github.com/swiftlang/swift-testing.git",
          branch: "main"),
],
```

Next, add the test target to the `targets` block using the following code:

```
testTarget(
  name: "MyProjectTests",
  dependencies: [
    "MyProject",
```

```
        .product(name: "Testing", package: "swift-testing"),
    ]
)
```

Now that we have seen how to add testing targets to our projects, let's explore Swift Testing in more detail.

Building blocks for Swift Testing

To effectively use Swift Testing, there are several key building blocks to understand. These components form the foundation of Swift Testing, enabling us to write, organize, and execute our tests efficiently.

We'll begin by looking at the @Test function.

Declaring the @Test function

The @Test attribute is a key component of Swift Testing, used to indicate that a function is a test, enabling Xcode to recognize it and display a **Run** button alongside it. This feature ensures that test functions are easily identifiable and executable within the Xcode environment.

Test functions with @Test can be functions or methods within a type, providing flexibility in how tests are organized and structured. These test functions can also be marked as async or throws, enabling asynchronous operations and error handling within our tests. They can also be isolated to a global actor if needed to ensure thread-safe execution.

The @Test attribute supports traits for specifying additional information on a per-test or per-suite basis. This capability allows for fine-tuning and customizing test behaviors and environments. The @Test attribute also facilitates parameterized testing by accepting arguments, which causes the test function to be called repeatedly for each argument. This simplifies the process of testing multiple input scenarios and ensures complete test coverage.

The following code shows how we would use the @Test attribute:

```
@Test func myTest() {
    // Test code here
}
```

When we create tests within Xcode, an empty diamond appears next to the test function, as shown in the following screenshot. We will see how to use this in the next section.

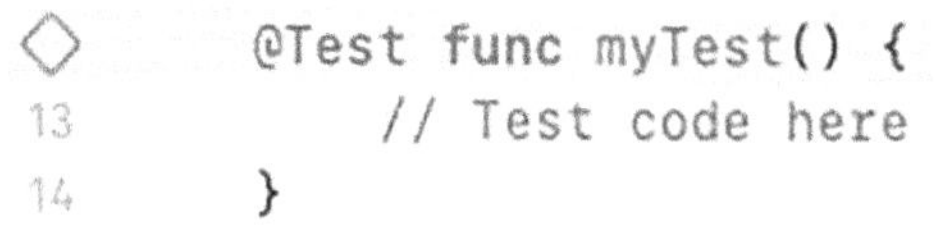

```
◇          @Test func myTest() {
13              // Test code here
14          }
```

Figure 18.5: Diamond for test case

Now let's look at the next building block, which is expectations.

Expectations

The #expect macro in Swift Testing is used for expressing expectations and assertions in test cases. It is the primary method for validating conditions, simplifying the process of writing tests by replacing the numerous assertion functions found within XCTest. This macro also enables us to express expectations as Boolean expressions, offering greater flexibility and simplicity compared to other testing frameworks. Another key advantage of the macro is its ability to capture sub-expression values, providing full diagnostic information when a test fails.

The #expect macro can be used for various types of assertions, including equality checks, error handling, and custom conditions. It also supports different forms of error handling, such as verifying that the code doesn't throw an error or that it throws a specific error.

The syntax of the macro is designed to feel natural for Swift developers, leveraging the language's expression capabilities. This ensures that writing and reading tests are straightforward and align with the overall Swift programming experience.

The following code shows how we would use the #expect macro in a very basic example:

```
@Test func validExpectation() async throws {
    #expect(1 == 1)
}
```

When a test function is defined with an #expect macro within the Xcode environment, we can click on the diamond next to the function to run the tests. Once the tests are run, either a green check or a red **x** will appear in the diamond, indicating whether the test passed or failed.

The following screenshot illustrates this.

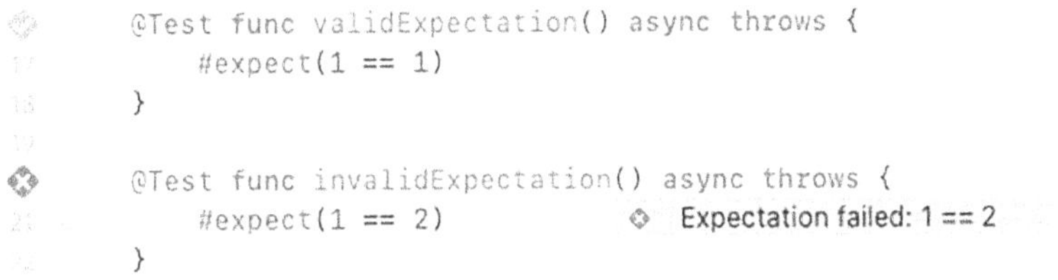

```
    @Test func validExpectation() async throws {
        #expect(1 == 1)
    }

    @Test func invalidExpectation() async throws {
        #expect(1 == 2)                    ◇ Expectation failed: 1 == 2
    }
```

Figure 18.6: Showing the results of test cases

Another macro that can be used is the #require macro.

#require macro

The #require macro in Swift Testing is primarily used to unwrap optional values and ensure that they are not nil within the test. When using this macro, if an optional value is nil, it will immediately fail the test.

Test functions that utilize #require must be marked as throwing, and the try keyword must precede the macro. This allows the macro to throw an error if the optional value is nil. Let's look at an example of this:

```
let one: Int? = 10
let two: String? = nil

let willSucceed = try #require(one)
let willFail = try #require(two)
```

In this example, the #require(one) code will pass and unwrap the optional one constant because it is not nil; however, the #require(two) code will fail because two is nil, causing the test to end prematurely.

While the #expect macro is used for expressing anticipated results, the #require macro focuses on conditions that must be satisfied for the test to proceed. This distinction aids in writing more precise tests. Now, let's look at confirmations with Swift Testing.

Confirmations

Swift 6.1 introduced a powerful enhancement to Swift Testing with range-based count confirmations using the `confirmation(expectedCount:)` function. This feature is especially useful for verifying that a certain number of items are produced, processed, or returned in a test case.

Instead of manually checking counts or writing custom error messages, Swift enables us to state our expectations clearly and in an easy-to-read format. If the count doesn't meet the expected value or range, the test will fail with a very precise and easy-to-read message.

Here's how confirmation works:

```
await confirmation(expectedCount: 5) { confirm in
    // Call confirm() exactly five times for success
}
```

Or, for a range of values:

```
await confirmation(expectedCount: 3...6) { confirm in
    // Call confirm() between 3 and 6 times for success
}
```

If the number of `confirm()` calls falls outside the specified range or is not equal to the expected count, Swift produces an error similar to this:

```
"Expected count to be within 3...6, but got 7."
```

Let's look at an example of how to use this. This test generates a random array and confirms that its count falls within a specific range:

```
@Test
func testRandomArrayCountInRange() async throws {
    let range = 5...10
    let count = Int.random(in: 0...15)
    let values = Array(repeating: "Item", count: count)

    await confirmation(expectedCount: range) { confirm in
        for _ in values {
            confirm()
        }
    }
}
```

This test case defines a valid range (5...10) for how many items should be present in the array. It generates a random number (0...15) to simulate variability or edge cases and then creates an array with that many items in it. Finally, for each item in the array, it calls `confirm()`, exactly once per item. If the number of confirmations (i.e., items) is outside the specified range, the test fails with a message indicating what was expected and what was observed.

Now, let's look at a new feature of Swift Testing in Swift 6.2, exit tests.

Exit tests

One of the challenges in Swift Testing has been the inability to verify code that triggers critical failures, the kind that terminate your process entirely, such as `preconditionFailure` or `fatalError`. With Swift 6.2, that limitation is addressed with exit tests, introduced as part of ST-0008. This feature enables us to safely and predictably test code that is expected to crash, by executing it in a subprocess and checking whether the process exited as expected.

In order to test code that results in a crash or forced termination, ST-0008 defines the `#expect(processExitsWith:)` macro. This macro runs the provided code in a dedicated subprocess, which isolates it from the main test runner and evaluates how it exits. You can specify that the process should exit with either `.success` (normal termination) or `.failure` (crash or assertion failure).

For this to work, the test function must be marked as async, and the #expect block must be awaited. This is to allow the test to be suspended while Swift launches and monitors the subprocess in the background.

Let's take a look at how this works. In this self-contained example, we demonstrate the use of an exit test to verify a crash caused by dividing by 0:

```
@Test
func testDivisionByZeroTriggersFailure() async throws {
    await #expect(processExitsWith: .failure) {
        let numerator = 42
        let denominator = Int.random(in:0...1)

        precondition(denominator != 0, "Cannot divide by zero")
        let _ = numerator / denominator
    }
}
```

This test defines two integers, where the denominator is a random number of either 0 or 1. A precondition is defined that ensures that division by 0 is not allowed. If the condition fails, Swift triggers a trap that would normally crash the test runner. By using #expect(processExitsWith: .failure), Swift now runs the code in a subprocess and confirms that it fails exactly as expected.

If the precondition were removed or changed, and the division was somehow allowed to proceed, the test would fail, not due to a crash, but because the process didn't exit with the expected failure.

A couple of things to note about exit tests:

- Only one #expect(processExitsWith:) call is allowed per test.
- Subprocesses do not share state with the main test runner—global variables, mocks, or side effects won't carry over.
- Tests that expect crashes must always use await since subprocess handling is asynchronous.

Now, let's look at the next building block, traits.

Traits

Traits are a very useful feature of Swift Testing that enables us to add detailed metadata and control the conditions under which tests are executed. By adding metadata to our tests, we can add things such as a display name, bug ticket reference, or other relevant information, improving our documentation and making it easier to understand the purpose and context of each test. Traits also enable conditional execution by specifying the conditions under which a test should run. For example, we can enable a test to run only when a certain feature flag is active or when running in a specific environment.

Traits can also help organize our tests into tags, which can be particularly useful for grouping tests by certain criteria, such as functionality. Tags can enable the filtering and running of specific subsets of tests within Xcode, making test management easier. They also simplify the creation of parameterized tests, enabling a single test to run multiple times with different parameters. By defining arguments outside the test, each test that is run can be displayed within the Xcode sidebar, with the ability to rerun the specific arguments as needed.

Here are a few examples of how traits can be used:

```
//Issue in bug tracker
@Test(.bug("FML1234", "A Bug"))

//Custom Tag
@Test(.tag(.critical))
```

```
//Conditionally enable or Unconditionally disable test
@Test(.enabled(if: required.isTrue))
@Test(.disabled("Test Broken"))

//Maximum time for test
@Test(.timeLimit(.minutes(2)))
```

Now that we have seen how traits are used, let's look at the final building block of Swift Testing, suites.

Suites

A suite can be created by embedding tests within a structure, and any structure containing @Test functions is automatically considered a suite. We can also create a suite using the @Suite annotation. Suites can be nested within other suites, allowing for a hierarchical structure for better test organization. Suites are very useful for organizing our tests, especially when dealing with hundreds of tests.

Suites can be structures, actors, or classes, but structures are encouraged due to their value semantics. Traits can be applied to suites, which will then be inherited by all tests within that suite. As an example, applying a tag to a suite will automatically apply that tag to all tests within the suite.

Let's look at two examples demonstrating how to define suites:

```
struct Project_Test {

    @Test func myTest() {
        // Test code here
    }
}


@Suite("Suite Example")
struct Project_Test {

    @Test func myTest() {
        // Test code here
    }
}
```

Both of these examples are considered suites because any structure containing @Test functions is automatically considered a suite.

Now that we have covered the basic building blocks of Swift Testing, let's examine an example of how to use it.

Swift Testing example

To illustrate how to use Swift Testing in our applications, let's consider a practical scenario: creating a calculator application.

> When creating a new project with Swift Testing, note that UI tests using XCTest are also created.

Creating our calculator

A calculator is a straightforward yet effective example for demonstrating testing concepts, as it involves basic arithmetic operations that can be easily verified with basic tests. Below, you'll find an example of the backend code for our calculator application:

```swift
struct Calculator {
    static func add(_ one: Double, _ two: Double) -> Double {
        one + two
    }

    static func subtract(_ one: Double, _ two: Double) -> Double {
        one - two
    }

    static func multiply(_ one: Double, _ two: Double) -> Double {
        one * two
    }
    static func divide(_ one: Double, _ two: Double) -> Double {
        one / two
    }
}
```

When we create the calculator application with Swift Testing, it creates three modules: one module for the application itself, one for unit tests, and one for UI tests. This code would go into the module for the application itself, which is the Calculator module, as shown in the following screenshot.

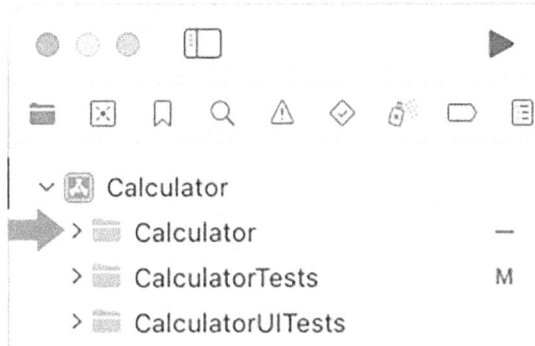

Figure 18.7: The Calculator module

Now, when we add unit tests, they go into the CalculatorTests module, shown in this screenshot:

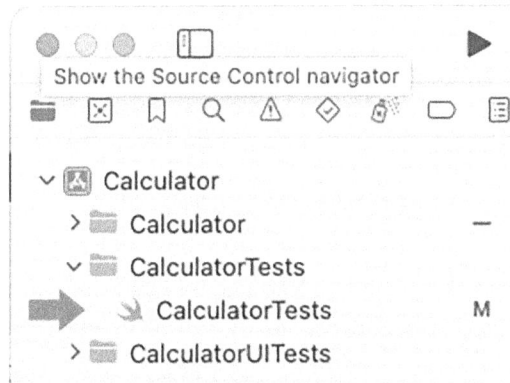

Figure 18.8: The CalculatorTests module

With Swift's access controls, which we discussed in *Chapter 17*, we know that when we create a type such as the Calculator structure without any explicit access levels defined, it defaults to the internal access level. What this means is that the Calculator type can be used by any entity within the same module. However, since our unit tests are defined in a different module, they will not have access to this type by default. To address this, Swift provides the @testable attribute, which enables our unit tests to access internal entities of the module. Let's take a look at this attribute.

Using the @testable attribute

The @testable attribute in Swift is a powerful tool that enables us to import a module with elevated access levels. This allows our unit tests, located in a different module, to test the internal components of our application code. This is particularly useful for accessing internal properties and methods that are not publicly exposed to code outside of the module they are defined in.

To use this attribute, we need to ensure that our module is compiled with testability enabled. This is achieved by setting the **Enable Testability** build setting to **Yes** for our application target. This configuration allows the compiler to generate the necessary metadata to support the @testable import. The following screenshot shows this setting:

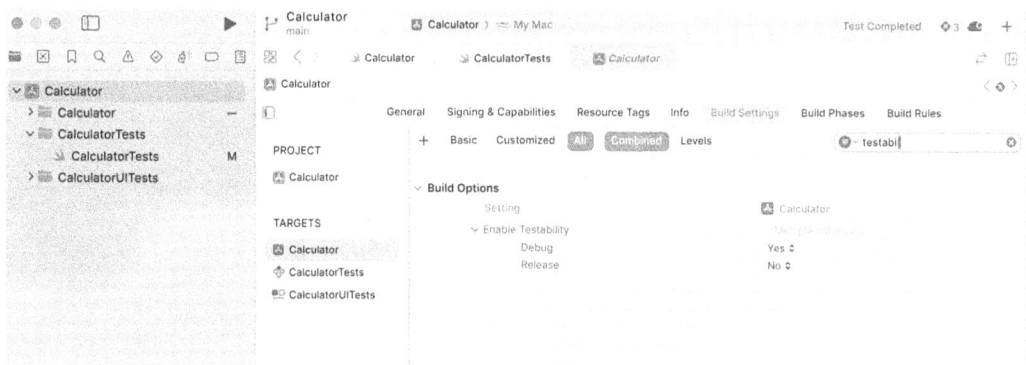

Figure 18.9: Enable Testability setting

💡 **Quick tip:** Need to see a high-resolution version of this image? Open this book in the next-gen Packt Reader or view it in the PDF/ePub copy.

📖 **The next-gen Packt Reader** and a **free PDF/ePub copy** of this book are included with your purchase. Scan the QR code OR visit packtpub.com/unlock, then use the search bar to find this book by name. Double-check the edition shown to make sure you get the right one.

After we have set the **Enable Testability** setting, to use @testable in our calculator application, we will want to add the following line to the top of our CalculatorTests.swift file:

```
@testable import Calculator
```

We are now able to access the internal types, methods, and properties of the Calculator module within our unit tests.

While the @testable attribute allows access to internal members, private properties and methods remain inaccessible. However, you can expose private properties for testing using the private(set) modifier, which allows read access from outside the type. The following code shows how we would use this:

```
struct SomeType {
    private(set) var someProperty = 42
}
```

Now that we can access the code within the Calculator module in our unit tests, we are ready to define some tests.

Testing our calculator

When developing tests with Swift Testing, it's best to create suites in order to organize our unit tests. Let's start by creating a testing suite called CalculatorTest, as shown in the following code:

```
@Suite("Calculator test")
struct Calculator_Test {

}
```

This code creates a suite for organizing our unit tests that will validate the calculator's functionality.

Next, let's add a unit test to this suite that will verify the functionality within the add() function. The following code shows one way to do this:

```
@Test func simpleAdditionTest() {
    #expect(Calculator.add(2, 2) == 4)
}
```

In this test function, we verify the add( ) function of the Calculator type by adding two numbers and checking the result. While this test would pass, there is a potential issue. What if, when we created our add( ) function, we accidentally introduced a typo, resulting in the following code?

```
static func add(_ one: Double, _ two: Double) -> Double {
        one * two
    }
```

The simpleAdditionTest unit test would still pass because 2 multiplied by 2 equals 4. Surprisingly, issues like this are common in unit tests. However, with Swift Testing, we can use arguments to run the tests multiple times with different values. Let's see how we can achieve this. First, we need to create a type in the CalculatorTests module to hold the arguments. This type will contain the following code:

```
struct TestValues {
    let first: Double
    let second: Double
    let answer: Double
}
```

This is a very basic type that holds three values: the two numbers for the math operation and then the expected result. Using this type, we can rewrite our addition unit test to test with multiple values, like this:

```
@Test("Addition Tests", arguments: [
    TestValues(first: 2, second: 3, answer: 5),
    TestValues(first: 10, second: 11, answer: 21),
    TestValues(first: 3.5, second: 4.5, answer: 8)
])
func testAddition(_ values: TestValues) async throws {
    #expect(Calculator.add(values.first, values.second) == values.answer)
}
```

In this code, we add a couple of traits to the unit test. The first trait is a name, "Addition Tests", and the second is an array of TestValues types, which defines the values to test with. We will note that the test function accepts a single argument of the TestValues type. Within the test function, we use these values to verify the calculator's add( ) function. When we run this test, it will verify that the add( ) function is working correctly.

Now let's add additional unit tests for our subtraction, multiplication, and division functions. These unit tests would look like this:

```
@Test("Subtraction Tests", arguments: [
        TestValues(first: 2, second: 3, answer: -1),
        TestValues(first: 11, second: 10, answer: 1),
        TestValues(first: 5, second: 4.5, answer: 0.5)
    ])
    func testSubtraction(_ values: TestValues) async throws {
        #expect(Calculator.subtract(
                values.first, values.second) == values.answer)
    }


    @Test("Multiplication Tests", arguments: [
        TestValues(first: 2, second: 3, answer: 6),
        TestValues(first: 11, second: 10, answer: 110),
        TestValues(first: 5, second: 4.5, answer: 22.5)
    ])
    func testMultiply(_ values: TestValues) async throws {
        #expect(Calculator.multiply(
                values.first, values.second) == values.answer)
    }


    @Test("Division Tests", arguments: [
        TestValues(first: 6, second: 3, answer: 2),
        TestValues(first: 11, second: 1, answer: 11),
        TestValues(first: 20, second: 5, answer: 4)
    ])
    func testDivide(_ values: TestValues) async throws {
        #expect(Calculator.divide(
                values.first, values.second) == values.answer)
    }
```

These unit tests are similar to our addition unit test, where we use the `arguments` trait to define multiple test values. Since we grouped all of the tests in a suite, we have the ability to run all of the tests by clicking on the diamond by the suite name; running these unit tests in Xcode makes it easy to see the results as well. The following screenshot illustrates how we can view the outcomes of our tests:

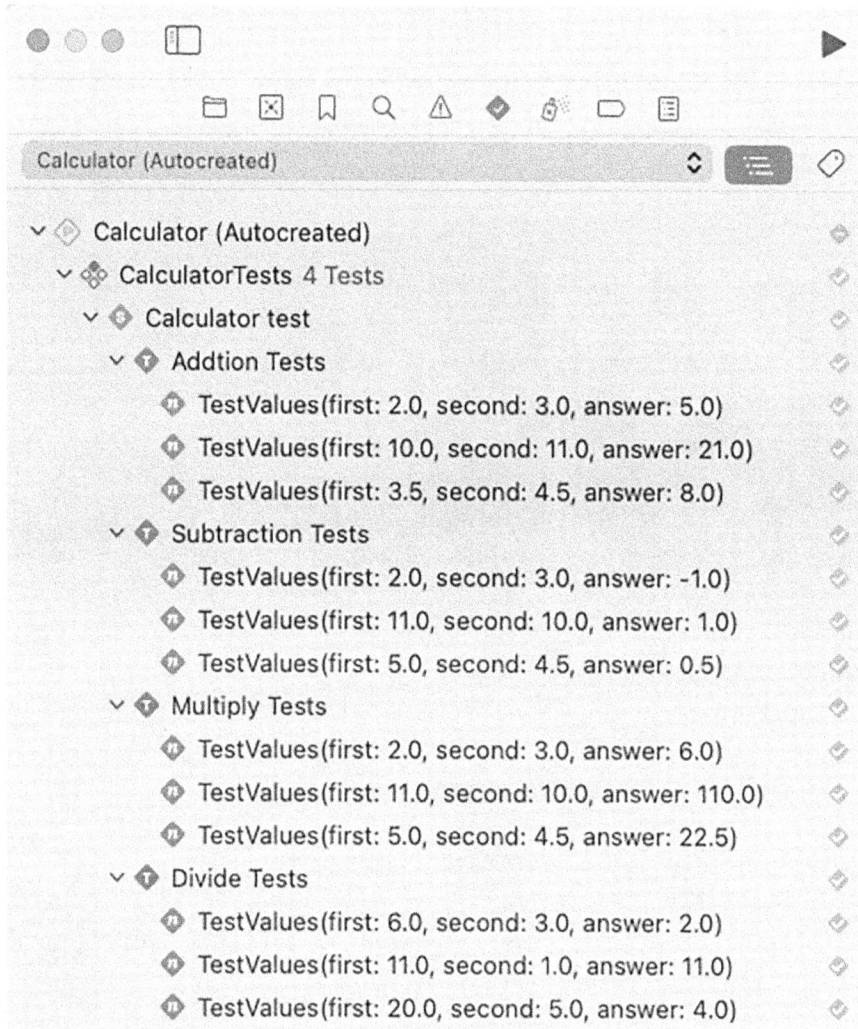

Figure 18.10: Viewing the outcomes of our tests

Within Xcode's left sidebar, if we click on the triangle icon with the checkmark in it, we can see that all of the unit tests passed. Let's change one of the test values to see what it looks like when a test fails. The following screenshot shows what the sidebar looks like when a test fails:

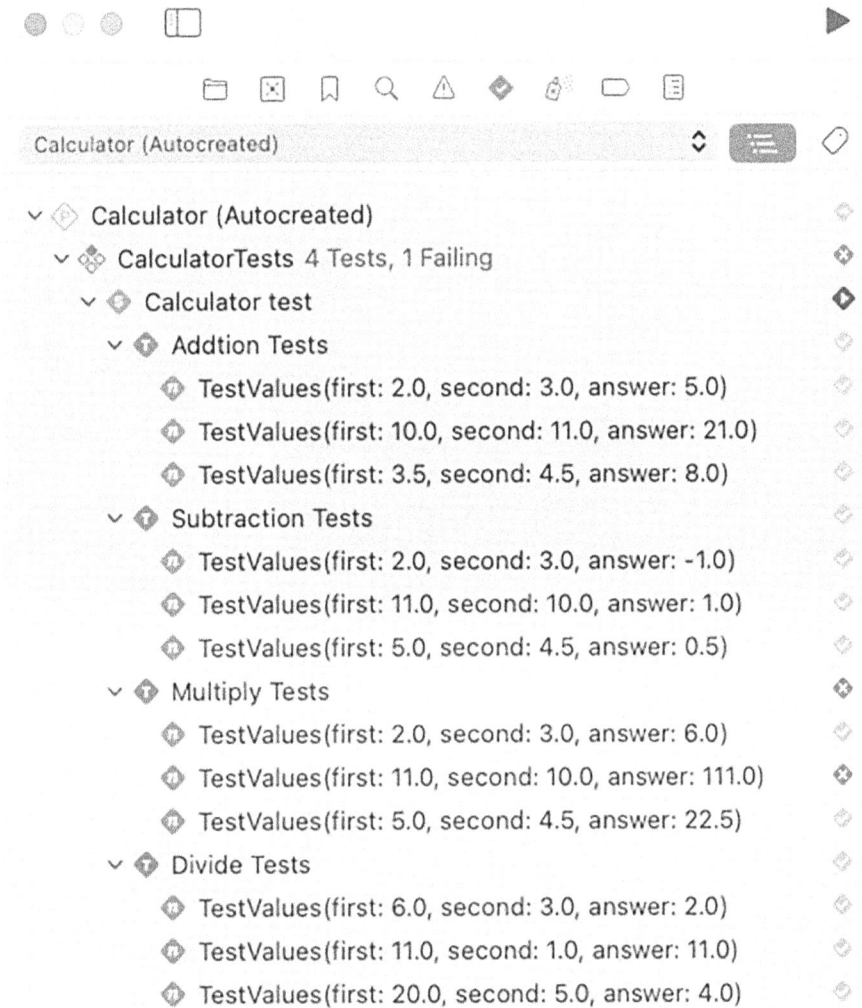

Calculator (Autocreated)

ν ⬡ Calculator (Autocreated)
  ν ⬢ CalculatorTests 4 Tests, 1 Failing
    ν ⬢ Calculator test
      ν ⬢ Addtion Tests
          ⬢ TestValues(first: 2.0, second: 3.0, answer: 5.0)
          ⬢ TestValues(first: 10.0, second: 11.0, answer: 21.0)
          ⬢ TestValues(first: 3.5, second: 4.5, answer: 8.0)
      ν ⬢ Subtraction Tests
          ⬢ TestValues(first: 2.0, second: 3.0, answer: -1.0)
          ⬢ TestValues(first: 11.0, second: 10.0, answer: 1.0)
          ⬢ TestValues(first: 5.0, second: 4.5, answer: 0.5)
      ν ⬢ Multiply Tests
          ⬢ TestValues(first: 2.0, second: 3.0, answer: 6.0)
          ⬢ TestValues(first: 11.0, second: 10.0, answer: 111.0)
          ⬢ TestValues(first: 5.0, second: 4.5, answer: 22.5)
      ν ⬢ Divide Tests
          ⬢ TestValues(first: 6.0, second: 3.0, answer: 2.0)
          ⬢ TestValues(first: 11.0, second: 1.0, answer: 11.0)
          ⬢ TestValues(first: 20.0, second: 5.0, answer: 4.0)

Figure 18.11: Identifying test failures

Xcode will also give us a lot of details on why the tests failed. To see these details, we can click on the failed test on the sidebar to take us to the area of the code that failed, and then click on the error to show the details. The following screenshot shows an example of this:

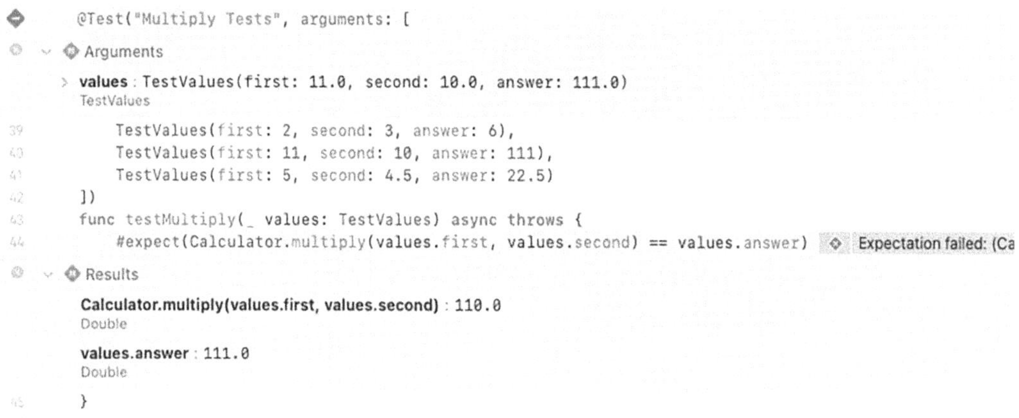

```
◆        @Test("Multiply Tests", arguments: [
  ◇  ⌄  ◆ Arguments
        >  values : TestValues(first: 11.0, second: 10.0, answer: 111.0)
           TestValues
39              TestValues(first: 2, second: 3, answer: 6),
40              TestValues(first: 11, second: 10, answer: 111),
41              TestValues(first: 5, second: 4.5, answer: 22.5)
42         ])
43         func testMultiply(_ values: TestValues) async throws {
44             #expect(Calculator.multiply(values.first, values.second) == values.answer)   ◇  Expectation failed: (Ca
  ◇  ⌄  ◆ Results
           Calculator.multiply(values.first, values.second) : 110.0
           Double

           values.answer : 111.0
           Double
45         }
```

Figure 18.12: Exploring why the tests failed

In the screenshot, we can see that the failing argument is listed at the top, while the calculated and expected values are shown at the bottom. This level of detail makes it very easy to identify the cause of issues within our unit tests. This raises the question of what to do if we have a known issue that will cause our unit tests to fail, but we still wish to continue running the tests to verify other changes to our code. For this, we can use the withKnownIssue function. The following code demonstrates how to use this:

```
@Test func simpleAdditionTest() {
    withKnownIssue("Addition will fail") {
        #expect(Calculator.add(2, 2) == 5)
    }
}
```

By using the withKnownIssue function, when we run our unit tests, a gray **x** will appear next to the test instead of a red **x**, indicating that there is a known issue with the test. The following screenshot illustrates this:

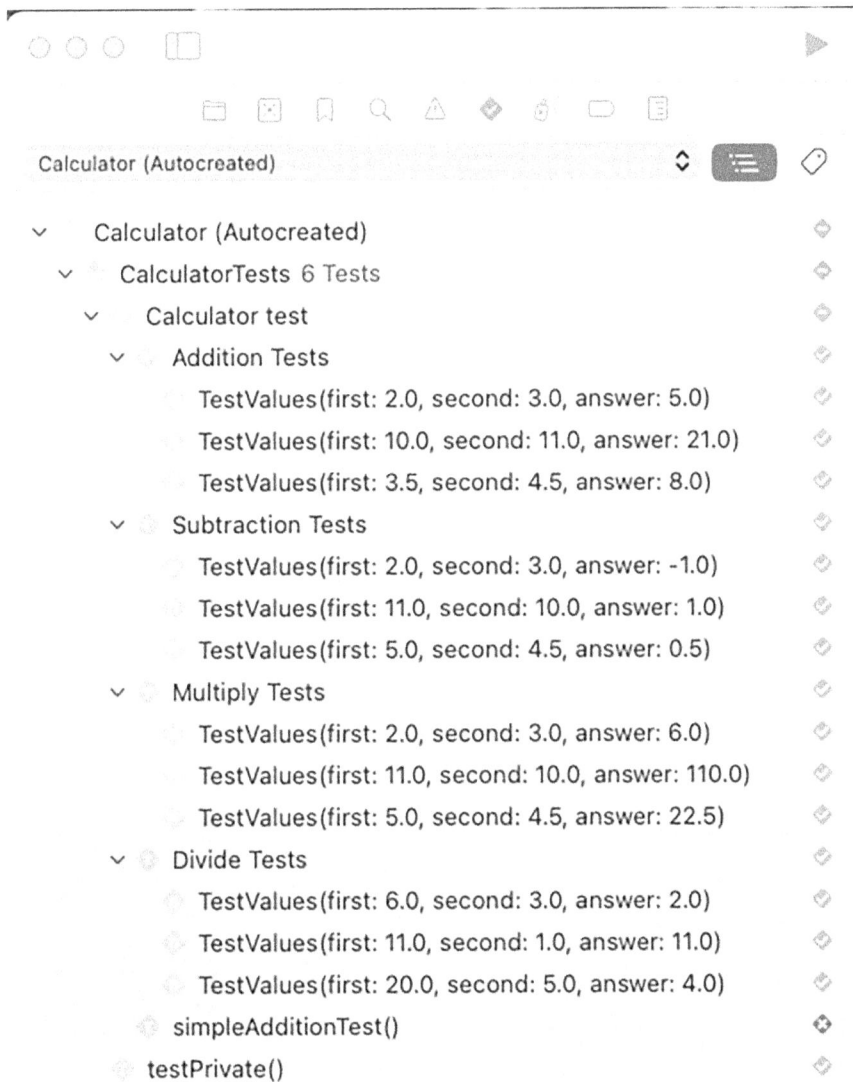

Calculator (Autocreated)

∨   Calculator (Autocreated)
  ∨   CalculatorTests 6 Tests
    ∨   Calculator test
      ∨   Addition Tests
          TestValues(first: 2.0, second: 3.0, answer: 5.0)
          TestValues(first: 10.0, second: 11.0, answer: 21.0)
          TestValues(first: 3.5, second: 4.5, answer: 8.0)
      ∨   Subtraction Tests
          TestValues(first: 2.0, second: 3.0, answer: -1.0)
          TestValues(first: 11.0, second: 10.0, answer: 1.0)
          TestValues(first: 5.0, second: 4.5, answer: 0.5)
      ∨   Multiply Tests
          TestValues(first: 2.0, second: 3.0, answer: 6.0)
          TestValues(first: 11.0, second: 10.0, answer: 110.0)
          TestValues(first: 5.0, second: 4.5, answer: 22.5)
      ∨   Divide Tests
          TestValues(first: 6.0, second: 3.0, answer: 2.0)
          TestValues(first: 11.0, second: 1.0, answer: 11.0)
          TestValues(first: 20.0, second: 5.0, answer: 4.0)
      simpleAdditionTest()
    testPrivate()

Figure 18.13: Using the withKnownIssue function

An additional feature of this function is that when the code is fixed and the test passes again, a red **x** will appear next to the test, prompting us to check it. The error will indicate that the test has passed, reminding us to remove the `withKnownIssue` function from the code so the unit test can be run normally.

Summary

At WWDC 2024, one of the most exciting tools unveiled was Swift Testing, which makes testing Swift code more efficient than ever. One of the key features is that Swift Testing and the older XCTest can coexist within the same test target and bundle. This allows us, over time, to convert our existing XCTest-based tests to Swift Testing, ensuring a smooth transition.

In this chapter, we discussed multiple methods for adding Swift Testing targets to our projects, including both Xcode projects and those created with the Swift Package Manager. We also discussed the key building blocks of Swift Testing, including the `@Test` attribute, which marks test functions, and the `#expect` and `#require` macros, which are used for assertions and optional unwrapping, respectively. Traits were covered, to add metadata and control test execution conditions, enhancing the organization and flexibility of test management, and suites were introduced to organize and group related tests, making it easier to manage large test collections. We ended this chapter with a practical example of testing a calculator application, demonstrating how to apply the concepts discussed.

In the next chapter, let's start looking at how we can architect our projects starting with one of the most popular programming paradigms, object-oriented programming.

19

Object-Oriented Programming

Object-Oriented Programming (OOP) is a widely adopted programming paradigm that organizes software around data, or objects, rather than focusing solely on functions and logic. This approach encourages the creation of reusable and modular components that represent real-world entities or concepts. In OOP, objects can store data and perform actions, making it easier to design complex systems in a more intuitive and structured manner. By aligning software design with real-world objects and actions, OOP principles can simplify the development and maintenance of large-scale applications.

OOP is widely used in many modern programming languages, including Swift.

In this chapter, we will learn:

- How Swift can be used as an OOP language
- How we can develop a class hierarchy in an object-oriented way
- What the benefits of OOP are
- What the drawbacks of OOP are

Before we see how Swift can be used as an OOP language, let's get a better understanding of what OOP is.

What is OOP?

OOP is a design philosophy that is fundamentally different from the older procedural programming approach used in languages like C and Pascal. While procedural programming relies on a series of instructions, or procedures, to tell the computer what to do, OOP focuses on objects, which encapsulate both data and behavior.

An object in OOP is a data structure that contains both attributes (properties) and actions (methods). An object can represent both real-world and virtual entities. For example, everyday objects like trees, grass, dogs, and fences can all be modeled with properties and actions.

Consider an example of a can of Jolt Cola, an energy drink that was very popular when I was in college. This could be modeled as an object with attributes like volume, caffeine content, temperature, and size. It could also have actions like drinking and changing temperature. Similarly, a cooler can be another object, with attributes like current temperature, how many cans of Jolt Cola it holds, and its maximum capacity, along with actions like adding and removing cans.

At the core of OOP is understanding how objects interact. For instance, placing a can of Jolt Cola in a cooler with ice will make that can colder. However, if there's no ice, the can will stay at its current temperature. These interactions must be considered when designing objects as they influence how the objects are designed.

In computer applications, objects are created from blueprints called classes. A class is a construct that allows us to encapsulate the properties and actions of an object into a single type, modeling the entity we are representing in our code. Using initializers within classes, we can create instances of the class, setting initial property values and performing any necessary setup.

Classes are the foundation of OOP. They are reference types and can have hierarchical relationships through inheritance, where one class can have a single superclass and many subclasses.

At the core of OOP are four key principles: encapsulation, inheritance, polymorphism, and abstraction. Encapsulation involves grouping data and methods that operate on the data into a single unit, while restricting access to certain components, which helps to protect the integrity of the data. Inheritance allows a class to inherit properties and methods from a parent class, promoting code reuse and creating a natural hierarchy. Polymorphism lets objects be treated as instances of their parent class, rather than their actual class, allowing more flexible and dynamic code. Abstraction simplifies complex systems by focusing on essential characteristics, while hiding unnecessary details from the user.

To illustrate these OOP concepts, let's design vehicle types for a video game. We will define the requirements for these vehicle types and show how to implement them using OOP principles. In the following chapter, we will redesign these classes using a protocol-oriented approach, providing a comparative understanding of both paradigms.

Requirements for the example code

When we develop applications, we usually have a set of requirements that we need to work toward. Our example project in this chapter is no different. The following is a list of requirements for the vehicle types within the game that we are creating:

- Our design will have three categories of vehicle: sea, land, and air. A vehicle may be a member of multiple categories.

- Vehicles may move or attack when they are on a tile that matches any of the categories they are in.

- Vehicles will be unable to move to or attack on a tile that does not match any of the categories they are in.

- When a vehicle's hit points reach 0, the vehicle will be considered incapacitated. We will also need to keep all active vehicles in a single array that we can loop through so that we can loop through them when they are moving or attacking.

In this chapter, we will demonstrate our design using a limited number of vehicles, knowing that the number of vehicle types will expand as we continue developing our game. Our primary focus will be on the design aspects; therefore, we won't be implementing much of the logic that governs the vehicles' movements and attacks.

Let's begin designing our vehicles in an object-oriented way.

Using object-oriented design

Swift fully supports developing applications using object-oriented principles. When it was first released, it was primarily viewed as an object-oriented language, much like Java and C#.

In this section, we will design the vehicle types using an object-oriented approach and examine the advantages and disadvantages of this design.

Before looking at the code, let's create a basic diagram to illustrate the vehicle class hierarchy and group related classes in our object-oriented design. With Swift being a single-inheritance language, each class can inherit from only one superclass. The root class in the hierarchy is the only class without a superclass.

Visualizing our class hierarchy

I typically begin with a simple diagram that outlines the classes without much detail. This allows me to visualize the class hierarchy. The following diagram shows the class hierarchy for our object-oriented design:

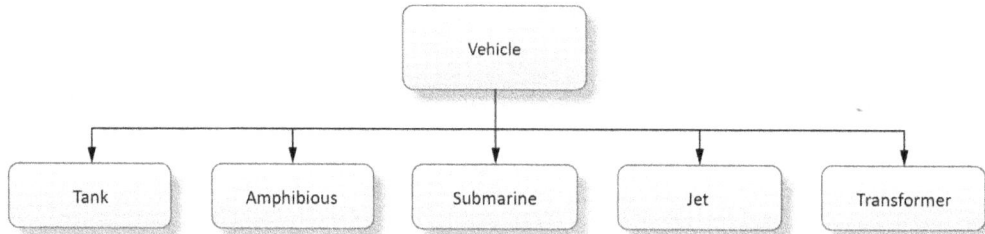

Figure 19.1: Vehicle class hierarchy

This diagram shows that we have one superclass named Vehicle and five subclasses: Tank, Amphibious, Submarine, Jet, and Transformer. In a class hierarchy, each subclass inherits all the properties and methods from the superclass. Therefore, any common code and properties can be implemented within the Vehicle superclass, and all subclasses will inherit it.

Given our requirements, which include three categories (land, air, and sea) of vehicles, we might consider creating a more extensive class hierarchy with superclasses defined for land, air, and sea vehicles. This would allow us to separate the code for each category into its own superclass. However, this approach is not feasible due to the nature of our requirements because a vehicle type may belong to multiple categories (land, air, and sea). With Swift supporting only single inheritance, each class can have only a single superclass. For example, if we created separate superclasses for land and sea, then either the land or the sea type could be the superclass to the Amphibious type, but not both.

The following figure illustrates this limitation:

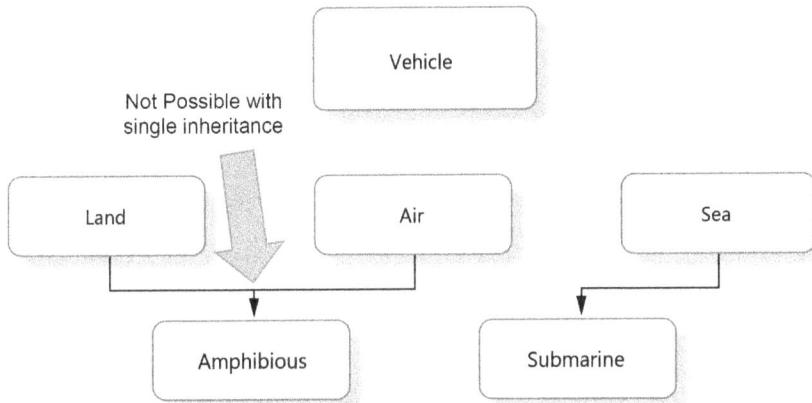

Figure 19.2: Limits of single inheritance

This limitation means that the Vehicle superclass will need to contain the code required for each of the three categories of vehicles. Having a single superclass such as this is one of the drawbacks of object-oriented design because the superclass can become very bloated.

Forming our object-oriented design

We will start forming our object-oriented design by creating a TerrainType enumeration that will be used to define the different vehicle attack and movement types. The TerrainType enumeration is defined like this:

```
enum  TerrainType {
    case  land
    case  sea
    case  air
}
```

Now, let's look at how we define the Vehicle superclass and the properties within this class:

```
class  Vehicle  {
    fileprivate var vehicleTypes  =  [TerrainType]()
    fileprivate var vehicleAttackTypes  =  [TerrainType]()
    fileprivate var vehicleMovementTypes  =  [TerrainType]()
    fileprivate var landAttackRange  =  -1
    fileprivate var seaAttackRange = -1
    fileprivate var airAttackRange = -1
    fileprivate var hitPoints  =  0
}
```

We begin defining the Vehicle type by specifying seven properties. The first three properties are arrays of the TerrainType type. These arrays track the vehicle type (the vehicleTypes array), the types of terrain from which the vehicle can attack (the vehicleAttackTypes array), and the types of terrain to which the vehicle can move (vehicleMovementTypesarray).

The next three properties (landAttackRange, seaAttackRange, and airAttackRange) represent the vehicle's attack range for each terrain type. An attack range with a value of less than 0 indicates that the vehicle cannot perform that particular type of attack. Finally, the last property, hitPoints, will track the vehicle's health.

Ideally, each of these properties, except for hitPoints, should be constants. However, subclasses cannot modify the value of a constant defined in a superclass; therefore, to manage access to these properties, we will use Swift's access controls, as shown in *Chapter 17*.

In this code, we defined the properties as fileprivate variables in order to restrict access to only subclasses while preventing external entities from modifying them. This access control allows properties and methods to be accessed by any code within the same source file. This requires that we create the subclass in the same source file as the superclass, which is not ideal because the file could become very large. However, in this object-oriented design, it is the best option to prevent these properties from being altered by instances of other types. If we have more than a few vehicle types, we might consider changing the access control to internal to allow the implementation of vehicles in separate files.

Since the properties are marked as `fileprivate`, we need to create getter methods to retrieve their values. Additionally, we will create methods to determine the types of terrain from which the vehicle can attack and move. Let's look at these methods:

```
func  isVehicleType(type:  TerrainType)  ->  Bool  {
    return  vehicleTypes.contains(type)
}
func  canVehicleAttack(type:  TerrainType)  ->  Bool  {
    return  vehicleAttackTypes.contains(type)
}
func  canVehicleMove(type:  TerrainType)  ->  Bool  {
    return  vehicleMovementTypes.contains(type)
}
func  doLandAttack()  {}
func  doLandMovement()  {}

func  doSeaAttack()  {}
func  doSeaMovement()  {}

func  doAirAttack()  {}
func  doAirMovement()  {}

func  takeHit(amount:  Int)  {  hitPoints  -=  amount  }
func  hitPointsRemaining()  ->Int  {  return  hitPoints  }
func  isAlive()  ->  Bool  {  hitPoints  >  0  }
```

The `isVehicleType` method accepts a parameter of the `TerrainType` type and returns true if the `vehicleTypes` array contains that terrain type. This enables external code to check if the vehicle is of a certain type. The next two methods also accept a `TerrainType` parameter and return true if the `vehicleAttackTypes` or `vehicleMovementTypes` array contains that terrain type. These methods are used to determine whether a vehicle can move to or attack from a certain type of terrain.

Following these methods are six additional methods that define attacks to or movement from different terrains for the vehicle. The next two methods are used to deduct hit points when the vehicle takes damage and to return the remaining hit points. The final method checks if the vehicle is still alive. However, there are some immediate issues with this design that we should consider.

The first issue, and one we have already briefly mentioned, is using the fileprivate access control and preventing direct access to the properties. This means that all subclasses need to be in the same physical source file as the Vehicle superclass. Given the potential size of the vehicle classes, consolidating them in one source file may not be the best option. To prevent this, we could set the property's access controls to internal; however, this would not prevent the properties from being altered by instances of other types within this package.

Another issue is that we need to provide methods for the vehicle to attack and move across each terrain type, even though most vehicles won't be able to operate in all three terrain types. Even though there is no code in the method implementations, external code will still be able to call any of the attack and movement methods. For instance, even though the Submarine type is a sea-only vehicle, external code could still call the land and air attack and movement methods.

Bloated superclasses, like our Vehicle type, pose a challenge in single-inheritance OOP languages like Swift. Such superclasses make it easy to mistakenly grant functionality that a type shouldn't have or deny it functionality it should have. For example, it would be easy to incorrectly set the airAttackRange property for the Submarine type, giving it the ability to attack from the air, which a submarine obviously cannot do.

> In this example, we are defining only a small subset of the functionality required for our vehicle types in a video game. Think about how large the Vehicle superclass could become if we implemented all the necessary functionality.

Let's look at how we would subclass the Vehicle class by creating the Tank, Amphibious, and Transformer classes. We will start with the Tank class:

```swift
class Tank: Vehicle {
    override init() {
        super.init()
        vehicleTypes = [.land]

        vehicleAttackTypes = [.land]
        vehicleMovementTypes = [.land]
        landAttackRange = 5
        hitPoints = 68
    }

    override func doLandAttack() {
```

```
            print("Tank  Attack")
        }
        override  func  doLandMovement()  {
            print("Tank  Move")
        }
    }
}
```

The Tank class is a subclass of the Vehicle class, and we start by overriding the default initializer in order to define the properties for this type. In this initializer, we assign values to several inherited properties, adding "land" to the vehicleTypes, vehicleAttackTypes, and vehicleMovementTypes arrays. This specifies that the Tank is a land vehicle capable of attacking and moving on land tiles. However, using arrays to track the vehicle type and its terrain capabilities can introduce errors. Even experienced developers might inadvertently add incorrect values, leading to unexpected behavior.

In the Tank class, we also override the doLandAttack() and doLandMovement() methods from the Vehicle superclass, since the Tank is a land vehicle. We don't override other attack and movement methods from the Vehicle superclass because the Tank isn't supposed to move or attack from the sea or air. Despite not overriding these methods, they remain part of the Tank class due to inheritance, leaving them accessible to external code.

Next, let's look at the Amphibious and Transformer classes, which are similar to the Tank class but can operate across multiple types of terrain. We'll first explore the Amphibious class, which can operate on both land and sea:

```
class  Amphibious:  Vehicle  {
    override  init()  {
        super.init()
        vehicleTypes  =  [.land,  .sea]
        vehicleAttackTypes  =  [.land,  .sea]
        vehicleMovementTypes  =  [.land,  .sea]

        landAttackRange  =  1
        seaAttackRange  =  1

        hitPoints  =  25
    }
    override  func  doLandAttack()  {
        print("Amphibious  Land  Attack")
```

```
    }
    override  func  doLandMovement()  {
        print("Amphibious  Land  Move")
    }
    override  func  doSeaAttack()  {
        print("Amphibious  Sea  Attack")
    }
    override  func  doSeaMovement()  {
        print("Amphibious  Sea  Move")
    }
}
```

The `Amphibious` class is similar to the `Tank` class we previously looked at. The main difference is that while the `Tank` type is defined for land-only operations, the `Amphibious` type is designed to operate on both land and sea. As a result, we override the attack and movement methods for both land and sea. Additionally, we include both "sea" and "land" values in the `vehicleTypes`, `vehicleAttackTypes`, and `vehicleMovementTypes` arrays.

Now, let's look at the `Transformer` class. This class is capable of operating across all three terrain types (land, sea, and air):

```
class  Transformer:  Vehicle  {
    override  init()  {
        super.init()
        vehicleTypes  =  [.land,  .sea,  .air]
        vehicleAttackTypes  =  [.land,  .sea,  .air]
        vehicleMovementTypes  =  [.land,  .sea,  .air]

        landAttackRange = 7
        seaAttackRange = 10
        airAttackRange = 12

        hitPoints  =  75
    }

    override  func  doLandAttack()  {
        print("Transformer  Land  Attack")
    }
```

```
    override func doLandMovement() {
        print("Transformer Land Move")
    }

    override func doSeaAttack() {
        print("Transformer Sea Attack")
    }
    override func doSeaMovement() {
        print("Transformer Sea Move")
    }

    override func doAirAttack() {
        print("Transformer Air Attack")
    }
    override func doAirMovement() {
        print("Transformer Air Move")
    }
}
```

For the Transformer type, we override all three movement and attack methods from the `Vehicle` superclass because the `Transformer` can operate on land, sea, and air. We also include all three terrain types in the `vehicleTypes`, `vehicleAttackTypes`, and `vehicleMovementTypes` arrays.

Using our vehicle types

Now that we have created these different vehicle types, let's look at how they can be used. One of our initial requirements was to maintain instances of all vehicle types within a single collection. This enables us to loop through all active vehicles and perform necessary actions. This task is made possible by polymorphism.

Polymorphism, derived from the Greek words "poly" (many) and "morph" (forms), refers, in computer science, to using a single interface to interact with multiple types within our code. This enables us to interact with different types uniformly. In OOP languages, polymorphism is achieved through a class hierarchy, enabling us to interact with various subclasses using the interface provided by a superclass.

Let's see how we can use polymorphism to manage instances of the different vehicle types in a single array. Since all vehicle types are subclasses of the Vehicle superclass, we can create an array of vehicle types and store instances of any subclass of Vehicle, as demonstrated below:

```
var  vehicles  =  [Vehicle]()

var vh1 = Amphibious()
var vh2 = Amphibious()
var vh3 = Tank()
var vh4 = Transformer()

vehicles.append(vh1)
vehicles.append(vh2)
vehicles.append(vh3)
vehicles.append(vh4)
```

Now, we can loop through and interact with each instance with the interface presented by the Vehicle type. The following code illustrates this:

```
for  (index,  vehicle)  in  vehicles.enumerated()  {
    if  vehicle.isVehicleType(type:  .air)  {
        print("Vehicle  at  \(index)  is  Air")
        if  vehicle.canVehicleAttack(type:  .air)  {
            vehicle.doAirAttack()
        }

        if  vehicle.canVehicleMove(type:  .air)  {
            vehicle.doAirMovement()
        }
    }

    if  vehicle.isVehicleType(type:  .land){
        print("Vehicle  at  \(index)  is  Land")
        if  vehicle.canVehicleAttack(type:  .land)  {
            vehicle.doLandAttack()
        }
        if  vehicle.canVehicleMove(type:  .land)  {
            vehicle.doLandMovement()
```

```
            }
        }
        if  vehicle.isVehicleType(type:  .sea)  {
            print("Vehicle  at  \(index)  is  Sea")
            if  vehicle.canVehicleAttack(type:  .sea)  {
                vehicle.doSeaAttack()
            }
            if  vehicle.canVehicleMove(type:  .sea)  {
                vehicle.doSeaMovement()
            }
        }
    }
```

In this code, we loop through the vehicles array and use the isVehicleType(type:) method to determine if a vehicle belongs to a specific type. We then call the appropriate movement and attack methods. We don't use an if-else or switch statement here because a vehicle may belong to multiple types, and we need to check each type even if the vehicle matched a previous type.

If we need to filter the results to only include vehicle instances of a specific type, we can use a where clause with the for loop. The following code demonstrates this approach:

```
    for  (index,  vehicle) in  vehicles.enumerated()  where vehicle.
    isVehicleType(type:  .air)  {
        if  vehicle.isVehicleType(type:  .air)  {
            print("**Vehicle  at  \(index)  is  Air")
            if  vehicle.canVehicleAttack(type:  .air)  {
                vehicle.doAirAttack()
            }

            if  vehicle.canVehicleMove(type:  .air)  {
                vehicle.doAirMovement()
            }
        }
    }
```

This code would only perform the attack and movement methods if the isVehicleType(type:) method returned true for the air type.

This design works well enough, and it shows how we would design our types using an object-oriented approach. However, as we will see in *Chapter 20, Protocol-Oriented Design*, we are able to resolve a lot of the issues presented here with a protocol-oriented design. Let's review the drawbacks of object-oriented design so we can see how protocol-oriented programming addresses them in the next chapter.

Issues with the object-oriented design

Two of the issues we encountered with object-oriented design are directly related to Swift being a single-inheritance language, meaning that a class can have only one superclass.

In a single-inheritance language like Swift, object-oriented design can result in bloated superclasses because we may need to include functionality that only a few subclasses require. This leads to the second issue of subclasses inheriting unnecessary functionality that they don't use.

In our design, we had to include functionality for all three terrain types because the vehicle types might need to operate on any of the terrains. This extra functionality could lead to errors if we are not careful. It's easy to accidentally create a class similar to this:

```swift
class Infantry: Vehicle {
    override init() {
        super.init()
        vehicleTypes = [.land]
        vehicleAttackTypes = [.land]
        vehicleMovementTypes = [.sea]

        landAttackRange = 1
        seaAttackRange = 1

        hitPoints = 25
    }
    override func doLandAttack() {
        print("Amphibious Land Attack")
    }
    override func doLandMovement() {
        print("Amphibious Land Move")
    }
}
```

Looking at this code, it's clear that the `vehicleMovementTypes` array incorrectly contains the sea type instead of the land type. Mistakes like this are easy to make.

Another issue with object-oriented design is the inability to create constants in the superclass that can be set by subclasses. In our design, there are several properties we would like to initialize in the subclasses and then never change. Ideally, these would be constants, but a constant defined in a superclass cannot be set in a subclass.

The final issue is the inability to restrict access to a property or method solely to subclasses. To work around this, we used the `fileprivate` access control so only code within the same source file could access the properties. However, this isn't an ideal solution because we may not want to put all of the subclasses in the same source file as the superclass. If we placed the subclasses in separate files, we would then have to set the access controls to internal, which would not prevent other types within the project from modifying them.

Summary

OOP is a design philosophy that is fundamentally different from the older procedural programming approach. While procedural programming relies on a set of instructions or procedures, OOP focuses on objects that encapsulate both data (attributes) and behavior (methods). These objects can represent real-world or virtual entities, allowing a more intuitive and modular approach to software design. In this chapter, we used the example of how a can of Jolt Cola can be modeled as an object with properties such as volume and caffeine content, and actions like drinking and temperature changes.

In OOP, objects are created from blueprints called classes, which define their properties and actions. Classes can form hierarchical relationships through inheritance, where subclasses inherit the characteristics of their superclasses. This promotes code reuse and organizes related functionality together. However, in Swift, a class can have only one superclass, leading to potential issues such as bloated superclasses, which contain functionality that some subclasses do not need.

In this chapter, we illustrated OOP design principles by creating vehicle types for a video. We began the design with a very basic class hierarchy, where a `Vehicle` superclass contains common properties and methods, and subclasses like `Tank`, `Amphibious`, and `Transformer` inherit from it. Despite the advantages of this approach, such as clear organization and code reuse, we saw that there are also significant drawbacks. The limitations of single inheritance in Swift highlight the challenges of maintaining clean and efficient code in OOP, setting the stage for exploring an alternative approach like protocol-oriented programming in the next chapter.

Unlock this book's exclusive benefits now

Scan this QR code or go to packtpub.com/unlock, then search this book by name.

Note: Keep your purchase invoice ready before you start.

20

Protocol-Oriented Programming

Protocol-Oriented Programming (POP) is a software design paradigm that focuses on using protocols to define methods, properties, and other requirements for types. Unlike **Object-Oriented Programming (OOP)**, which relies on class inheritance, as we saw in *Chapter 19*, POP encourages the use of protocols to create flexible and reusable components.

One of the key advantages of POP is its ability to reduce the tight coupling that is often found in class hierarchies. By defining protocols, developers can specify a set of requirements that types must conform to, without dictating how these requirements should be implemented.

Swift's standard library is built using a protocol-oriented design, showcasing the effectiveness of this design philosophy. Many core types within Swift, such as collections and numeric types, conform to protocols that define their essential behaviors. This design allows for a high degree of flexibility and interoperability within the language.

In this chapter, we will learn about the following topics:

- What is POP?
- What is protocol inheritance?
- What is protocol composition?
- How to use protocol extensions

In *Chapter 19*, we explored how to design vehicle types for a video game using object-oriented design principles. In this chapter, we will redesign those same vehicle types using a protocol-oriented approach. This will enable us to see the differences between these two approaches.

Next to some of the more advanced topics discussed in previous chapters, the examples in this chapter may appear somewhat basic, or even as though they've taken you back a step. This is intentional and designed to help you start thinking in a protocol-oriented manner, easing the transition from the object-oriented mindset you may be accustomed to. Once you achieve this mindset shift, you can begin integrating the advanced concepts we've previously discussed.

What is POP?

POP, which was introduced with Swift version 2.0, is one of the most exciting features of Swift. This development paradigm encourages developers to think differently about how they structure their code. POP shifts the focus from the rigid class inheritance of OOP to a more flexible way of defining and sharing behaviors, which leads to cleaner, more adaptable code.

Traditionally, OOP has been the go-to programming paradigm for building complex systems, relying on classes and inheritance to share functionality. However, as projects grow, this class-based approach can lead to tangled hierarchies and duplicated code, as seen in *Chapter 19*, which is where POP comes in. Rather than building structures based on what something "is" through inheritance, POP lets us focus on what something "does." This is achieved by creating protocols, which is where the name POP comes from. Protocols act as blueprints for behavior that can be shared across multiple types, such as structures, enumerations, and classes.

Swift's support for protocols goes beyond what many developers are used to. By using protocol extensions, you can add default behaviors to types that conform to protocols, which means you can define a set of actions or features once and then reuse them where needed. This makes your code easier to read and reduces the need for repetitive or boilerplate code.

Before we look at what POP is, let's begin by reviewing the requirements for our vehicle types that we defined in *Chapter 19*.

Requirements for the sample code

When we develop applications, we usually have a set of requirements that we need to work toward. Our sample project in this chapter is no different. The following is a list of requirements for the vehicle types that we are creating:

- Our design will have three categories of vehicles: sea, land, and air. A vehicle may be a member of multiple categories.

- Vehicles may move or attack when they are on a tile that matches any of the categories they are in.

- Vehicles will be unable to move to or attack on a tile that does not match any of the categories they are in.

- When a vehicle's hit points reach zero, the vehicle will be considered incapacitated. We will also need to keep all active vehicles in a single array that we can loop through.

Note that these are the same as the requirements in the previous chapter.

As in the previous chapter, in this chapter, we will demonstrate our design using a limited number of vehicles, knowing that the number of vehicle types will expand as we continue developing the game. Our primary focus will be on the design aspects, so we won't be implementing much of the logic that governs the vehicles' movements and attacks.

Using POP

While Swift was initially introduced as an object-oriented language, similar to Java and C#, with Swift version 2.0, it became a protocol-oriented language. Traditionally, a vehicle hierarchy in OOP might be structured with a Vehicle superclass and specific subclasses such as Tank, Submarine, and Jet. However, with Swift's single inheritance, each class can inherit from only one superclass, which has its limitations.

POP offers a more flexible approach. Instead of building class hierarchies, we define behaviors through protocols. This shift enables us to compose functionality without complex inheritance chains, making our code modular, reusable, and easier to scale.

Before looking at the code, let's create a basic diagram to illustrate how our protocol-oriented design would look.

Visualizing our protocol-oriented design

Similar to the approach we took with object-oriented design, we will begin by creating a simple diagram to illustrate how to design the vehicle types in a protocol-oriented manner. Like the object-oriented diagram in *Chapter 19*, this will be a straightforward representation, focusing on the types themselves without going into the implementation details:

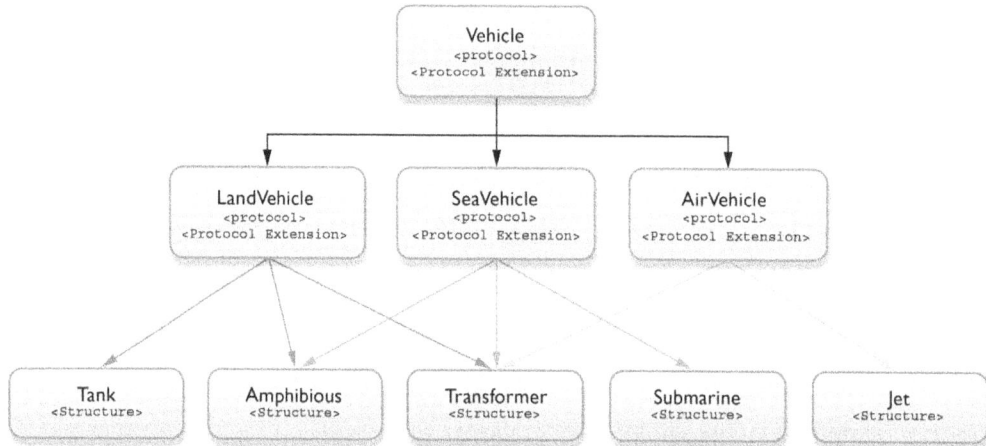

Figure 20.1: Protocol-oriented design

The protocol-oriented design is significantly different from the object-oriented design. In the object-oriented approach, we began with a superclass as the central focus, from which all sub-classes inherited functionality and properties.

In the protocol-oriented design, protocols and their extensions are the central focus. This new design employs three key techniques that distinguish it from OOP: protocol inheritance, protocol composition, and protocol extensions. Let's take a quick look at these three techniques.

Protocol inheritance enables one protocol to inherit the requirements from other protocols, similar to class inheritance in OOP. However, instead of inheriting functionality from a superclass, requirements are inherited from the protocol. For instance, in this example, the LandVehicle, SeaVehicle, and AirVehicle protocols all inherit the requirements from the Vehicle protocol. One significant advantage that protocol inheritance has over class hierarchies is that protocols can inherit requirements from multiple protocols.

It's important to note that by combining protocols with protocol extensions, our types can also inherit functionality. Protocol extensions enable us to provide default implementations for methods and properties defined within a protocol. What this means is that types conforming to a protocol automatically gain these implementations provided by the protocol extension, therefore inheriting functionality in a manner similar to traditional class inheritance. This combination of protocol inheritance and protocol extensions provides a powerful and flexible way to design and organize our code.

Protocol composition allows types to conform to more than one protocol, offering greater flexibility and modularity in our design. In our vehicle example, some types, such as the Tank, Submarine, and Jet structures, conform to a single protocol while other types, such as the Amphibious and Transformer structures, use protocol composition in order to conform to multiple protocols.

Protocol inheritance and composition are key elements of a protocol-oriented design because they enable us to create smaller, more specific protocols. This approach helps us avoid the bloated superclasses often seen with object-oriented designs. By breaking down functionality into distinct protocols, we can mix and match them as needed, creating more modular and reusable code. However, it's essential to strike a balance; creating protocols that are too granular can lead to difficulties in maintaining and managing the code base. While smaller, more specific protocols are beneficial, they should be thoughtfully designed to maintain a manageable level of complexity.

Protocol extensions allow us to extend a protocol by providing default implementations for conforming types, whether the function is defined within the protocol or not, thereby eliminating the need to provide an individual implementation for each type.

While protocol extensions may seem unremarkable at first, their true power becomes apparent with deeper understanding. They can fundamentally transform the way you approach application design once you truly understand them.

💡 **Quick tip**: Enhance your coding experience with the **AI Code Explainer** and **Quick Copy** features. Open this book in the next-gen Packt Reader. Click the **Copy** button (1) to quickly copy code into your coding environment, or click the **Explain** button (2) to get the AI assistant to explain a block of code to you.

```
                                                              Copy      Explain
function calculate(a, b) {
  return {sum: a + b};                                          1          2
};
```

📖 **The next-gen Packt Reader** is included for free with the purchase of this book. Scan the QR code OR go to packtpub.com/unlock, then use the search bar to find this book by name. Double-check the edition shown to make sure you get the right one.

Forming our protocol-oriented design

Now, let's begin implementing the vehicle types starting with the Vehicle protocol. In this example, the Vehicle protocol will define a single property called hitPoints, which will track the vehicle's remaining hit points:

```
protocol  Vehicle  {
    var  hitPoints:  Int  {get  set}
}
```

As you may recall from our object-oriented design in *Chapter 19*, there were three methods defined in the Vehicle superclass that were utilized by all vehicle types: takeHit(amount:), hitPointsRemaining(), and isAlive(). Since the implementation of these methods is identical for every vehicle type, they make ideal candidates for implementation in a protocol extension.

The following code shows how we could create an extension for the `Vehicle` protocol and implement these three methods within the extension:

```
extension Vehicle {
    mutating func takeHit(amount: Int) {
        hitPoints -= amount
    }
    func hitPointsRemaining() -> Int {
        return hitPoints
    }
    func isAlive() -> Bool {
        return hitPoints > 0 ? true : false
    }
}
```

Now, any type that conforms to the `Vehicle` protocol or conforms to any protocol that inherits from the `Vehicle` protocol will automatically receive these methods. When a protocol inherits requirements from another protocol, it also inherits the functionality provided by that protocol's extensions.

Next, let's look at how we would define the `LandVehicle`, `SeaVehicle`, and `AirVehicle` protocols for our design:

```
protocol LandVehicle: Vehicle {
    var landAttack: Bool {get}
    var landMovement: Bool {get}
    var landAttackRange: Int {get}

    func doLandAttack()
    func doLandMovement()
}

protocol SeaVehicle: Vehicle {
    var seaAttack: Bool {get}
    var seaMovement: Bool {get}
    var seaAttackRange: Int {get}

    func doSeaAttack()
    func doSeaMovement()
```

```
    }

    protocol  AirVehicle:  Vehicle  {
        var  airAttack:  Bool  {get}
        var  airMovement:  Bool  {get}
        var  airAttackRange:  Int  {get}


        func  doAirAttack()
        func  doAirMovement()
    }
```

There are a few important items to note about these protocols. Firstly, they all inherit the requirements from the Vehicle protocol, which means they also inherit the default functionality provided by the Vehicle protocol extension.

Additionally, these protocols contain only the relevant requirements for their specific vehicle types, unlike the Vehicle superclass in the object-oriented design, which included the requirements for all vehicle types. By dividing the requirements into three separate protocols, the code becomes safer, easier to maintain, and more modular. If common functionality is needed, we can simply add a protocol extension to one or more of the protocols.

Furthermore, the properties for these protocols are defined using only the get attribute, meaning they will be defined as constants within the conforming type. This is a significant advantage of the protocol-oriented design, as it prevents external code from changing the values once they are set.

Now, let's look at how to create types that conform to these protocols. We will implement the same Tank, Amphibious, and Transformer types as in the object-oriented design. Let's begin with the Tank type:

```
struct  Tank:  LandVehicle  {
    var  hitPoints  =  68
    let  landAttackRange  =  5
    let  landAttack  =  true
    let  landMovement  =  true


    func  doLandAttack()  {  print("Tank  Attack")  }
    func  doLandMovement()  {  print("Tank  Move")  }
}
```

There are several differences between the Tank type defined here and the one from *Chapter 19*. To highlight these differences, let's review the Tank type that was written using the object-oriented design paradigm:

```
class  Tank:  Vehicle  {
    override  init()  {
        super.init()
        vehicleTypes  =  [.land]

        vehicleAttackTypes  =  [.land]
        vehicleMovementTypes  =  [.land]
        landAttackRange  =  5
        hitPoints  =  68
    }

    override  func  doLandAttack()  {
        print("Tank  Attack")
    }
    override  func  doLandMovement()  {
        print("Tank  Move")
    }
```

The first noticeable difference is that the Tank type in our object-oriented design is a class, which is a reference type, while the Tank type in the protocol-oriented design is a structure, which is a value type. While protocol-oriented design doesn't mandate the use of value types, it is generally preferred. With that in mind, we could define the Tank type as a class in either paradigm. There may be times when it is necessary to use a reference type. However, value types should be preferred.

One of the main advantages of choosing value types over reference types is safety. With value types, a unique copy of an instance is always received, ensuring type safety. This is particularly useful in a multithreaded environment, where we don't want one thread to alter data while another thread is using it, as this can create difficult-to-replicate and hard-to-trace bugs. However, there are situations, as in our case, where we may need to make changes to vehicle instances and those changes must be persisted. Although this isn't typical behavior for a value type, we can achieve it using an inout parameter.

Another difference is that the Tank type in the protocol-oriented design uses the default initializer provided by the structure, allowing us to define the properties as constants. These constant properties, once set, can't be changed. In contrast, the Tank type from the object-oriented design required us to override the initializer and set the properties within it. The properties in the object-oriented design were also defined as variables, allowing them to be changed after they were set.

Finally, while not immediately obvious, the Tank type in the protocol-oriented design only includes functionality for land vehicles. The Tank type in the object-oriented design inherits functionality and properties for sea and air types as well as land types, even though it doesn't require that additional functionality.

Now let's see how we could create the Amphibious type:

```
struct  Amphibious:  LandVehicle,  SeaVehicle  {
    var  hitPoints  =  25
    let  landAttackRange  =  1
    let  seaAttackRange  =  1

    let  landAttack  =  true
    let  landMovement  =  true

    let  seaAttack  =  true
    let  seaMovement  =  true


    func  doLandAttack()  {
        print("Amphibious  Land  Attack")
    }
    func  doLandMovement()  {
        print("Amphibious  Land  Move")
    }
    func  doSeaAttack()  {
        print("Amphibious  Sea  Attack")
    }
    func  doSeaMovement()  {
        print("Amphibious  Sea  Move")
    }
}
```

The Amphibious type is similar to the Tank type but uses protocol composition to conform to both the LandVehicle and SeaVehicle protocols. This enables it to use functionality defined in both land and sea protocols.

Let's look at how we would implement the Transformer type:

```
struct  Transformer:  LandVehicle,  SeaVehicle,  AirVehicle  {
    var  hitPoints  =  75
    let  landAttackRange  =  7
    let  seaAttackRange  =  5
    let  airAttackRange  =  6

    let  landAttack  =  true
    let  landMovement  =  true

    let  seaAttack  =  true
    let  seaMovement  =  true

    let  airAttack  =  true
    let  airMovement  =  true

    func  doLandAttack()  {
        print("Transformer  Land  Attack")
    }
    func  doLandMovement()  {
        print("Transformer  Land  Move")
    }
    func  doSeaAttack()  {
        print("Transformer  Sea  Attack")
    }
    func  doSeaMovement()  {
        print("Transformer  Sea  Move")
    }
    func  doAirAttack()  {
        print("Transformer  air  Attack")
    }
    func  doAirMovement()  {
        print("Transformer  air  Move")
    }
}
```

Since the Transformer type can operate across all three terrain types, we use protocol composition to conform to the LandVehicle, SeaVehicle, and AirVehicle protocols.

Using our vehicle types

Now let's look at how we would use these new types. Similar to the object-oriented design, we need to keep instances of all vehicle types in a single collection. This enables us to loop through all active vehicles and perform necessary actions. Once again, we will use polymorphism for this. However, with the protocol-oriented design, we will use the interface provided by the protocols to interact with the vehicle instances.

We'll start by creating an array and adding several instances of the vehicle types to it:

```
var  vehicles  =  [Vehicle]()

var vh1 = Amphibious()
var vh2 = Amphibious()
var vh3 = Tank()
var vh4 = Transformer()

vehicles.append(vh1)
vehicles.append(vh2)
vehicles.append(vh3)
vehicles.append(vh4)
```

This code is similar to the code from the object-oriented design. Here, we create an array to store instances of types that conform to the Vehicle protocol. With protocol inheritance, this array will accept types conforming to protocols that inherit from the Vehicle protocol. In our example, this means the array can contain instances of types conforming to the LandVehicle, SeaVehicle, AirVehicle, and Vehicle protocols.

By defining the array to contain instances of types that conform to the Vehicle protocol, we can interact with these instances using the interface defined by the Vehicle protocol. Although the Vehicle protocol itself may not offer extensive functionality, we can typecast the instances to check whether they conform to a more specific protocol. The following code demonstrates this approach:

```
for  (index,  vehicle)  in  vehicles.enumerated()  {
    if  vehicle  is  AirVehicle  {
        print("Vehicle  at  \(index)  is  Air")
```

```
    }
    if  vehicle  is  LandVehicle  {
        print("Vehicle  at  \(index)  is  Land")
    }
    if  vehicle  is  SeaVehicle  {
        print("Vehicle  at  \(index)  is  Sea")
    }
}
```

In this code, we use a for loop to iterate through the vehicles array. We utilize the is operator to check whether the instances conform to one of the protocols (AirVehicle, LandVehicle, or SeaVehicle), and if they do, we print out a message.

Accessing the vehicle types in this way is quite similar to the object-oriented example. However, what if we only want to retrieve a specific type of vehicle rather than all vehicles? We can do this using the where clause. The following example demonstrates how to do this:

```
for  (_,  vehicle)  in  vehicles.enumerated()  where  vehicle  is
LandVehicle {
    let  vh  =  vehicle  as?  LandVehicle
    if  vh.landAttack  {
        vh.doLandAttack()
    }
    if  vh.landMovement  {
        vh.doLandMovement()
    }
}
```

In this example, we use the where keyword to filter the for loop results, retrieving only instances that conform to the LandVehicle protocol. We can then typecast any instance returned from the for loop as a LandVehicle and interact with it using the interface provided by the protocol.

Protocol inheritance and composition are very important to protocol-oriented design. Let's take a look at these key concepts in a little more detail.

Protocol inheritance

Protocol inheritance is a powerful feature that is part of almost any protocol-oriented design. Just as classes can inherit from another class, protocol inheritance allows a protocol to inherit the requirements of one or more other protocols. This creates an inheritance hierarchy where a child protocol inherits all requirements of its parent protocols, in addition to any new requirements it defines.

Unlike class inheritance, which is restricted to a single superclass, protocols can inherit from multiple protocols, allowing them to merge various sets of requirements, and promoting the creation of highly modular and reusable components.

By using protocol extensions, we can provide default implementations for the methods and properties defined by a protocol, and when a protocol inherits requirements from another protocol, it also inherits any default implementations provided by the extensions. This reduces the need for boilerplate code and can greatly enhance code maintainability.

Protocol composition

Protocol composition is a powerful feature that allows us to create types that conform to multiple protocols. This feature adds a greater level of flexibility and modularity to our code base, letting us build complex functionality by combining simple, more focused protocols.

Unlike class inheritance, where a class can only inherit from one superclass, protocol composition enables a type to conform to multiple protocols by inheriting the requirements and behaviors from each and combining requirements into a single, unified interface. This is done using simple, comma-separated syntax.

By defining small, focused protocols, we can mix and match them to create flexible types. This modular approach promotes code reuse and enhances code maintainability.

Protocol composition ensures that conforming types contain the requirements defined in each protocol, providing a clear contract for which methods and properties must be implemented. This ensures that all types adhere to the contracts set by the protocols they adopt.

Now let's look at another example to see how protocol inheritance, composition, and extensions can work together.

Protocol-oriented design

In protocol-oriented design, we emphasize the use of protocols to define the blueprint of methods, properties, and other requirements for types, promoting more flexible and modular code.

Within a protocol-oriented design, we start with the protocols. In this new example, we will begin by defining two protocols – the Nameable and Contactable protocols:

```
protocol Nameable {
    var firstName: String { get }
    var middleName: String? { get }
    var lastName: String { get }
}


protocol Contactable {
    var emailAddress: String { get }
    var phoneNumber: String { get }
}
```

Both of these protocols define properties that any type conforming to them must implement.

Now, we will use protocol inheritance to create a Person protocol that will inherit requirements from both of these protocols:

```
protocol Person: Nameable, Contactable {
    var birthDate: Date { get }
    var age: Int { get }
    func displayInfo()
}
```

By creating smaller, more modular protocols, we can reuse them, creating a tool chest of reusable components. For example, if we needed to create a Pet protocol for defining various pets that a person could own, we could reuse the Nameable protocol like this:

```
protocol Pet: Nameable {
    var numberOfLegs: Int { get }
}
```

Within the Person protocol, we define the birthDate and age properties. Knowing that age can be calculated from birth date, we can use a protocol extension to define this default functionality. The following code illustrates this:

```
extension Person {
    var age: Int {
        let now = Calendar.current
        let components = now.dateComponents([.year], from: birthdate,
```

```
                                                           to: Date.now)
        return components.year ?? 0
    }
}
```

Within this extension, we define a calculated property that calculates the age of a person based on their birth date. Now, any type that adopts the Person protocol or any type that adopts a protocol that inherits from the Person protocol will have this functionality. Before we create types that will conform to these protocols, let's create one more protocol, named Occupation:

```
protocol Occupation {
    var occupationName: String { get set }
    var yearlySalary:Double { get set }
    var experienceYears: Double { get set }
}
```

Within the Occupation protocol, we define a set of requirements that are used to define properties related to a person's job. Now, let's create an Employee type that will inherit the requirements from both the Person and Occupation protocols:

```
struct Employee: Person, Occupation {
    var firstName: String
    var middleName: String?
    var lastName: String
    var birthDate: Date
    var emailAddress: String
    var phoneNumber: String
    var occupationName: String
    var yearlySalary: Double
    var experienceYears: Double

    func displayInfo() {
        //Display Employee Information
    }

}
```

In the Employee type, we use protocol composition to ensure that it conforms to both the Person and Occupation protocols. Since the Person protocol inherits from the Nameable and Contactable protocols, the Employee type is also required to implement the properties and methods defined within these protocols. This means that an Employee type must implement the properties and methods specified by Nameable (firstName, middleName, and lastName), Contactable (emailAddress and phoneNumber), and Person (birthDate, age, and displayInfo), as well as those defined in the Occupation protocol (occupationName, yearlySalary, and experienceYears).

If we also needed to create a Consultant type, we could define it like this:

```
struct Consultant: Person {
    var firstName: String
    var middleName: String?
    var lastName: String
    var birthDate: Date
    var emailAddress: String
    var phoneNumber: String

    func displayInfo() {
        //Display Employee Information
    }
}
```

Since we do not need to store any occupation information for a consultant, it only needs to conform to the Person protocol. However, because the Person protocol inherits from the Nameable and Contactable protocols, the Consultant type must also implement the properties and methods defined in these protocols.

Additionally, within our application, we may want to define a Dog type that conforms to the Pet protocol. Where the Pet protocol inherits from the Nameable protocol, a Dog type may look like this:

```
struct Dog: Pet {
var numberOfLegs: Int { get }
    var firstName: String
    var middleName: String?
    var lastName: String
}
```

As demonstrated in our examples, using a protocol-oriented design with protocol composition, inheritance, and extensions enables us to create highly modular and reusable components. This approach results in a code base that is easy to expand and maintain over the long term.

Swift does not only promote a protocol-oriented design for our code bases but it is also how the Swift standard library is designed.

Swift standard library

If we explore the documentation for the Swift standard library at https://developer.apple.com/documentation/swift/swift-standard-library, it quickly becomes evident that the library is developed using a protocol-oriented design. As an example, if we scroll through the documentation for the integer type, we can see that it conforms to 34 protocols, including the following:

- BinaryInteger
- Comparable
- Decodable
- Encodable
- EntityIdentifierConvertible
- Equatable
- Numeric
- SignedInteger
- SignedNumeric

Many of the standard library types are built around protocols, providing a high degree of flexibility and interoperability. This design philosophy is evident in the extensive use of protocols such as Equatable, Comparable, and Collection, which define common interfaces and behaviors that multiple types adopt. By understanding and utilizing a protocol-oriented design, developers can make the most of Swift's capabilities.

Summary

In this chapter, we explored the fundamentals of POP in Swift to better understand the differences between that and OOP. Protocol-oriented design in Swift emphasizes the use of protocols and protocol extensions, unlike traditional OOP, which focuses on classes and class hierarchies.

We looked at the key techniques of POP, including protocol inheritance, protocol composition, and protocol extensions. Protocol inheritance allows one protocol to adopt the requirements of another, enabling the creation of more specific and focused protocols. Protocol composition lets types conform to multiple protocols, providing a high degree of flexibility and enabling the construction of complex functionalities from simpler components. Protocol extensions provide default implementations for protocol methods and properties, allowing conforming types to automatically receive functionalities without redundant code.

Finally, we highlighted how Swift's standard library is developed using a protocol-oriented approach. By examining the documentation, it became apparent that most of the standard library types are built around protocols, ensuring a high degree of flexibility and interoperability. This design philosophy, evident in protocols such as Equatable, Comparable, and Collection, underscores the power and effectiveness of a protocol-oriented design, encouraging developers to adopt these principles within their code base.

Now that we have seen how Swift works as both an object-oriented and protocol-oriented language, in the next chapter, let's look at how it can be used as a functional programming language.

Unlock this book's exclusive benefits now

Scan this QR code or go to packtpub.com/unlock, then search this book by name.

Note: Keep your purchase invoice ready before you start.

21

Functional Programming with Swift

Functional programming has become increasingly popular in recent years, offering developers a powerful paradigm for writing clean, maintainable, and efficient code. Swift embraces functional programming concepts alongside its object-oriented and protocol-oriented features. This versatility makes Swift an excellent choice for developers looking to explore and implement functional programming techniques in their projects.

At the heart of functional programming is the use of pure functions and immutability. Pure functions are those that always produce the same output for a given input, without modifying any external state. This predictability leads to code that is easy to test and debug.

Immutability is another key principle of functional programming, and Swift offers several ways of supporting immutable data. The language's `let` keyword for declaring constants, along with value types such as structs and enums, encourages developers to create immutable data structures. This approach helps prevent unexpected state changes and reduces the likelihood of bugs caused by mutable shared state in concurrent environments.

In addition to pure functions and immutability, Swift's support for first-class functions, higher-order functions, and closures provides a strong foundation for developers writing in a functional style.

In this chapter, we will learn about the following:

- What the core principles of functional programming are
- How to use functional programming techniques with Swift
- Some advanced functional programming techniques

Let's start off by looking at the core principles of functional programming.

Core principles of functional programming

Functional programming is a programming paradigm that views programs as collections of mathematical functions and avoids changing data or states. While Swift is primarily a protocol-oriented language, it offers support for functional programming concepts.

Functional programming is based on several fundamental principles, and in this section, we will explore five of the most important principles, starting with immutability.

Immutability

Immutability is a key concept of functional programming that helps make code more reliable and easier to understand. Basically, immutability means that once we create a piece of data, we can't or don't change it. Instead of modifying the original data, we create a new version of the data with the changes.

This is different from other programming paradigms where data can be changed or modified frequently, which can lead to unexpected errors and harder-to-track bugs. By preventing data from changing, we ensure that our code behaves in a predictable way, making it easier to track down bugs and other issues with our code.

Immutability also makes it safer to run parts of our program in a concurrent environment, because we don't have to worry about one part of our application changing data that another part is using. Overall, embracing immutability in functional programming leads to cleaner, more robust, and maintainable code.

Let's look at a code example to illustrate the concept of immutability with Swift:

```
let numbers = [1, 2, 3, 4, 5]
let doubled = numbers.map { $0 * 2 }
```

In this example, we begin by defining a constant that holds an array of numbers. Using constants instead of variables is critical for maintaining immutability because, once defined, a constant cannot be changed. When we want to double each number in the array, rather than updating the original array, a new constant is created that contains the new values.

Pure functions

Pure functions are a fundamental concept in functional programming, and they play a critical role in writing reliable and maintainable code. A pure function is one that, given the same input, always produces the same output and has no additional side effects. This means that pure functions do not modify any external state or depend on an external state, ensuring that their behavior is predictable and consistent.

The use of pure functions brings several advantages to functional programming. They make code easier to understand because each function operates independently of external factors. Pure functions also simplify testing and debugging since they can be tested in isolation without considering the broader application state. Furthermore, they enable safer concurrent execution, since the lack of side effects eliminates the risk of one part of the application interfering with another.

Let's look at a basic example of a pure function:

```
func add(_ first: Int, _ second: Int) -> Int {
    first + second
}

let total = add(2,4)
```

In this example, we define a function that takes two arguments and returns the sum of the values. The function operates solely on its input parameters, adding the two values together and returning the result without altering or relying on any external state. This makes the function a pure function, as it produces the same output given the same inputs and has no side effects.

First-class functions

First-class functions are another key concept in functional programming. They enable us to treat functions as first-class citizens, which means they can be assigned to variables, passed as arguments to other functions, and returned from other functions, just like other data types.

First-class functions enable support of key functional programming techniques such as currying and function composition. These techniques help us create a more modular and maintainable code base by breaking down complex operations into simpler, reusable components.

Let's look at a basic example of how a first-class function works. In this example, we will begin by creating the following two functions:

```swift
func add(_ first: UInt, _ second: UInt) -> UInt {
    first + second
}

func subtract(_ first: UInt, _ second: UInt) -> UInt {
    first - second
}
```

The first function, add(), will add two numbers together and return the results, while the second function, subtract(), will subtract the second number from the first and return the result. What is key to realize about these two functions is that they have the same function signature where they each accept two unsigned integers and return an unsigned integer ((UInt, UInt) -> UInt).

One of the concepts of first-class functions is the ability to assign them to variables or constants. The following example shows how we could do this:

```swift
let mathFunction = add
let result = mathFunction(8, 4)
```

In this example, we create a constant named mathFunction and assign the add() function to it. We then use the mathFunction constant to add two numbers together. Alternatively, we could have assigned the subtract() function to the mathFunction constant to subtract the numbers. If mathFunction were a variable instead of a constant, we could change which function is assigned to it at any time.

Higher-order functions

Higher-order functions are another key concept in functional programming and significantly improve the flexibility and reusability of our code. A higher-order function is one that can take other functions as arguments, return functions as the result, or both. This enables us to write more abstract and modular code by creating functions that operate on other functions.

In functional programming, higher-order functions enable powerful techniques such as function composition, where complex operations are built by combining simpler functions.

As an example, we will use the same add() or subtract() function that we created in the *First-class functions* section and create a function that accepts either of these two functions as an argument:

```
func performMathOperation(_ first: UInt, _ second: UInt,
                          function: (UInt, UInt) -> UInt) -> UInt {
    function(first, second)
}
```

In this example, notice that the third argument of the performMathOperation() function, named function, accepts a function with the same functional signature as the add() and subtract() functions, which is (UInt, UInt) -> UInt. Then, within this function, we call the function that was passed in, returning the result. We could then use the performMathOperation() function like this:

```
let result = performMathOperation(8, 4, function: subtract)
```

In this example, we call the performMathOperation() function, passing in the subtract() function as the third argument. After this code is run, the result constant will have a value of 4. (If we had passed in the add() function, the value of the result constant would have been 12.)

Swift includes several higher-order functions that are commonly used. Some examples of these are the map(_:), filter(_:), reduce(::), and forEach(_:) functions.

Now that we have seen some of the core principles of functional programming, let's look at some advanced functional programming techniques that Swift offers.

Advanced functional programming techniques

While Swift supports the core principles of functional programming, it also offers robust support for advanced techniques that enhance our ability to write code in a functional style. These advanced techniques include function composition, currying, and recursion, which enable us to create more modular, reusable, and expressive code. Let's look at these advanced techniques, starting with function composition.

Function composition

Function composition is a powerful concept in functional programming that involves combining two or more functions in order to create a new function. This new function is the result of the combined functions, executed in sequence. In Swift, function composition can be achieved using various techniques, including higher-order functions and operators.

At its core, function composition allows us to take the output of one function and use it as the input to another function. This chaining of functions can help create more readable and maintainable code by breaking down complex operations into simpler, reusable components.

Let's look at a very basic example where we have the following two functions:

```
func addOne(_ number: Int) -> Int {
    return number + 1
}

func toString(_ number: Int) -> String {
    String(number)
}
```

The first function accepts a single integer as the argument, adds 1 to it, and returns the resulting value. The second function also takes a single integer as the argument and returns the string value. Now, for very basic function composition, we could create a new function that combines these two functions like this:

```
func addOneToStringFunc(_ number: Int) -> String {
    toString(addOne(number))
}
```

This new function takes a single integer as the argument and then uses the addOne() function and the toString() function to add 1 to the number and return a string representation of the number.

Now let's look at how we can compose functions using an operator. The following code shows how to do this:

```
infix operator >>>

func >>> <A, B, C>(lhs: @escaping (A) -> B, rhs: @escaping (B) -> C) ->
                                                   (A) -> C {
    return { rhs(lhs($0)) }
}
```

This code begins by defining an infix operator, which is an operator that is placed between two operands. An example of the infix operator is the "+" or "-" operators that add two numbers or subtract two numbers. The custom operator is defined with three greater-than signs (>>>).

The next line implements the custom operator. It starts off by defining three generic type parameters: A, B, and C. We then define the left-hand (lhs) and right-hand (rhs) operands.

The left-hand operand is defined as @escaping (A) -> B, which means it takes a value of type A and returns a value of type B. The @escaping attribute is used to specify that the function can be stored and used after the function it is passed to has returned.

The right-hand operand is defined as @escaping (B) -> C, which means that it takes a value of type B and returns a value of type C. The @escaping attribute is also used for this operand.

The next part of this line, (A) -> C, defines that the return type for this operator is a function that takes a value of A, which is the value that the left-hand operand takes as a parameter, and returns a value of C, which is the value returned from the right-hand operand.

The body of this operator creates a closure that takes an input ($0), which will be a value of type A, and applies it to the left-hand operand (lhs($0)). The result of the left-hand operand is then applied to the right-hand operand.

> In this code, when we mention left-hand operands and right-hand operands, keep in mind that we are really talking about functions.

Now let's see how we can use this operator:

```
let addOneToString = addOne >>> toString
let result2 = addOneToString(3)
```

With this code, we use the >>> operator to compose the addone() and toString() functions. We then use the new addOneToString() function to add 1 to the parameter, which is the integer 3, and the result is a string representation of the number 4. While this approach may seem a bit complex for something as simple as our example here, the advantage is the ability to mix and match functions as needed. For example, let's say that we had another function that doubled the number passed in, as shown in this code:

```
func double(_ number: Int) -> Int {
    return number * 2
}
```

We could create another function very easily that took a value, doubled it, and returned the string representation, like this:

```
let doubleToString = double >>> toString
```

Using the operator method for function composition makes it very easy to mix and match various smaller functions to get the functionality that we need at the time. To mix and match functions in this way, the functions must have the same function signature.

Currying

Currying is a concept in functional programming that enables us to convert a function with multiple arguments into a sequence of functions, each with only a single argument. This enables functions to be broken down into smaller, more manageable pieces, improving code modularity and reusability.

In traditional function calls, a function may take multiple arguments and process them at once. For instance, this function adds two numbers:

```
func add(_ a: Int, _ b: Int) -> Int {
    return a + b
}


let result = add(2, 3)
```

The add() function is a pure function that takes two integers as parameters and then returns the sum of the two.

With currying, we can transform this function into a series of functions, like this:

```
func curriedAdd(_ a: Int) -> (Int) -> Int {
    return { a + $0 }
}


let addTwo = curriedAdd(2)
let result = addTwo(3)
```

The curriedAdd() function takes one integer as a parameter (func curriedAdd(_ a: Int)) and returns a function, which also takes one integer as a parameter and has a return type of an integer (-> (Int) -> Int). Within the function body, a closure is returned that adds the two numbers together.

In my experience, currying is a useful technique to understand but it's best used sparingly. While currying can improve the modularity and reusability of code by breaking down functions into smaller, composable pieces, overuse can lead to unnecessary complexity and reduced readability.

Recursion

Recursion is a concept in functional programming where a function calls itself to solve smaller parts of the same problem until the entire calculation is complete. In Swift, recursion can be used to elegantly solve problems that can be broken down into simpler, repeatable tasks. By using recursion, we can write expressive code that aligns with functional programming principles.

In a recursive function, the function calls itself with a subset of the original problem's input, gradually reducing the problem's size until it reaches its base case. The base case is a condition that stops the recursion, preventing an infinite loop.

Let's look at how we may calculate the factorial of a number using a recursive function. The following code shows how we may write this function:

```
func factorial(_ n: Int) -> Int {
    if n <= 1 {
        return 1
    } else {
        return n * factorial(n - 1)
    }
}
```

In this example, the factorial() function calls itself with the value of n-1 until n is less than or equal to 1. In this example, the base case is n <= 1, because that prevents this function from getting into an infinite loop.

Summary

Although Swift is primarily a protocol-oriented language, it also supports functional programming concepts for writing clean and efficient code. Core principles such as immutability, pure functions, first-class functions, and higher-order functions are built into the language itself. Immutability ensures that data cannot be altered once created, leading to predictable and reliable code. Pure functions consistently produce the same output given the same input without side effects from external states, simplifying testing. First-class functions, which can be assigned to variables, passed as arguments, or returned from other functions, enable advanced techniques such as currying and function composition, improving code modularity. Higher-order functions enhance flexibility and reusability by allowing functions to take other functions as arguments or return them as results. This enables us to write abstract and modular code-supporting techniques such as function composition.

Advanced functional programming techniques in Swift further enhance code flexibility. Function composition combines simple functions into complex operations, improving readability. Currying transforms multi-argument functions into sequences of single-argument functions, increasing modularity. However, currying should be used sparingly to avoid unnecessary complexity within our code. Recursion, where a function calls itself to solve smaller instances of a problem, provides elegant solutions for tasks such as calculating factorials. By integrating these principles and techniques into our code base, we can write robust, scalable, and maintainable applications.

22

Unlock Your Book's Exclusive Benefits

Your copy of this book comes with the following exclusive benefits:

- ☁ Next-gen Packt Reader
- ✦ AI assistant (beta)
- 📄 DRM-free PDF/ePub downloads

Use the following guide to unlock them if you haven't already. The process takes just a few minutes and needs to be done only once.

How to unlock these benefits in three easy steps

Step 1

Have your purchase invoice for this book ready, as you'll need it in *Step 3*. If you received a physical invoice, scan it on your phone and have it ready as either a PDF, JPG, or PNG.

For more help on finding your invoice, visit `https://www.packtpub.com/unlock-benefits/help`.

> **Note**: Did you buy this book directly from Packt? You don't need an invoice. After completing Step 2, you can jump straight to your exclusive content.

Step 2

Scan this QR code or go to packtpub.com/unlock.

On the page that opens (which will look similar to *Figure 22.1* if you're on desktop), search for this book by name. Make sure you select the correct edition.

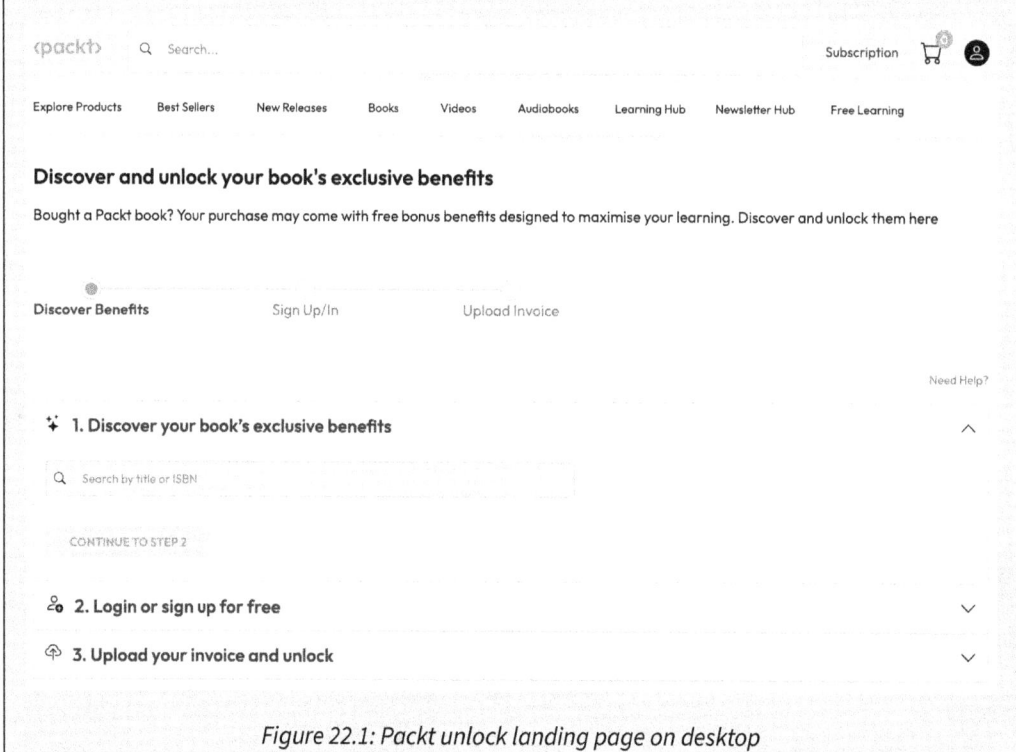

<packt>   Q Search...   Subscription 🛒 👤

Explore Products   Best Sellers   New Releases   Books   Videos   Audiobooks   Learning Hub   Newsletter Hub   Free Learning

Discover and unlock your book's exclusive benefits

Bought a Packt book? Your purchase may come with free bonus benefits designed to maximise your learning. Discover and unlock them here

Discover Benefits   Sign Up/In   Upload Invoice

                                                                                          Need Help?

✧ 1. Discover your book's exclusive benefits   ∧

   Q   Search by title or ISBN

   CONTINUE TO STEP 2

👤 2. Login or sign up for free   ∨

☁ 3. Upload your invoice and unlock   ∨

Figure 22.1: Packt unlock landing page on desktop

Step 3

Once you've selected your book, sign in to your Packt account or create a new one for free. Once you're logged in, upload your invoice. It can be in PDF, PNG, or JPG format and must be no larger than 10 MB. Follow the rest of the instructions on the screen to complete the process.

Need help?

If you get stuck and need help, visit `https://www.packtpub.com/unlock-benefits/help` for a detailed FAQ on how to find your invoices and more. The following QR code will take you to the help page directly:

Note: If you are still facing issues, reach out to `customercare@packt.com`.

‹packt›

packtpub.com

Subscribe to our online digital library for full access to over 7,000 books and videos, as well as industry leading tools to help you plan your personal development and advance your career. For more information, please visit our website.

Why subscribe?

- Spend less time learning and more time coding with practical eBooks and Videos from over 4,000 industry professionals

- Improve your learning with Skill Plans built especially for you

- Get a free eBook or video every month

- Fully searchable for easy access to vital information

- Copy and paste, print, and bookmark content

At www.packtpub.com, you can also read a collection of free technical articles, sign up for a range of free newsletters, and receive exclusive discounts and offers on Packt books and eBooks.

Other Books You May Enjoy

If you enjoyed this book, you may be interested in these other books by Packt:

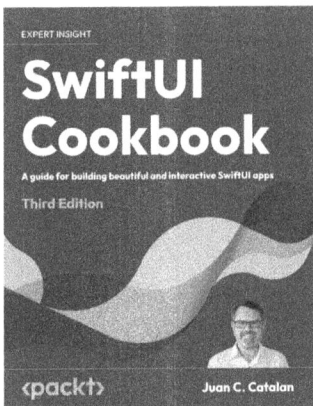

SwiftUI Cookbook, Third Edition

Juan C. Catalan

ISBN: 978-1-80512-173-2

- Create stunning, user-friendly apps for iOS 17, macOS 14, and watchOS 10 with SwiftUI 5
- Use the advanced preview capabilities of Xcode 15
- Use async/await to write concurrent and responsive code
- Create powerful data visualizations with Swift Charts
- Enhance user engagement with modern animations and transitions
- Implement user authentication using Firebase and Sign in with Apple
- Learn about advanced topics like custom modifiers, animations, and state management
- Build multi-platform apps with SwiftUI

iOS 18 Programming for Beginners, Ninth Edition

Ahmad Sahar

ISBN: 978-1-83620-488-6

- Learn the foundations of using Xcode 16 and Swift 6
- Implement the latest iOS 18 features through a hands-on example app
- Build responsive iOS apps using UIKit
- Create location-based apps using Core Location and MapKit
- Implement concurrency in Swift for asynchronous programming
- Build iOS apps using industry-standard design patterns and practices
- Enhance apps with Apple Intelligence to leverage machine learning
- Test apps with Swift Testing to ensure it meets quality standards

Packt is searching for authors like you

If you're interested in becoming an author for Packt, please visit authors.packt.com and apply today. We have worked with thousands of developers and tech professionals, just like you, to help them share their insight with the global tech community. You can make a general application, apply for a specific hot topic that we are recruiting an author for, or submit your own idea.

Share your thoughts

Now you've finished *Mastering Swift 6*, we'd love to hear your thoughts! Scan the QR code below to go straight to the Amazon review page for this book and share your feedback or leave a review on the site that you purchased it from.

https://packt.link/r/1836203691

Your review is important to us and the tech community and will help us make sure we're delivering excellent quality content.

Index

www.ingramcontent.com/pod-product-compliance
Lightning Source LLC
Chambersburg PA
CBHW081045220326
41598CB00038B/6988